Heft 657 (en)

DEUTSCHER AUSSCHUSS FÜR STAHLBETON

German Committee for Structural Concrete

Impact of Microstructural Changes in Concrete on Coda Waves

by

Dr.-Ing. Fabian Diewald

1st edition 2025

Herausgeber:
Deutscher Ausschuss für Stahlbeton e.V. – DAfStb

DIN Media GmbH

Summary

Monitoring the condition of concrete structures is a macrosocial task for safety and economic reasons that increasingly uses techniques from the field of non-destructive testing. A promising technique for structural health monitoring of heterogeneous materials such as concrete or reinforced concrete is coda wave interferometry, as it is sensitive to even weak changes in the material and can therefore be used as an early warning system for damage. The technique is based on comparing two ultrasonic wave forms to derive a change in velocity. The change is particularly well measurable and resolvable in heterogeneous media due to strong scattering in late arriving wave packages, the coda.

The objective of the study was to investigate the relationships between transient impacts, such as temperature, relative humidity, and external stress changes, on the velocity change of coda waves propagating in concrete. Furthermore, long-term phenomena such as shrinkage, creep, and hydration were characterized using the relative velocity change. The investigations published in three articles showed that temperature and stress changes in the elastic range are linearly related to the velocity change of the coda. The interaction of temperature and humidity, on the other hand, led to a non-linear change. Furthermore, the results demonstrated that shrinkage, creep, and hydration can be described not only by conventional measurement techniques but also by means of coda wave interferometry.

When using coda wave interferometry, it is crucial to be able to separate the individual effects to distinguish a reversible velocity change from irreversible damage. An approach for this separation is provided summarizing the possible impacts. All relationships were developed using a uniform concrete mix, with the objective that the results can be transferred to a multiscale model. Such a model can form the basis for the transferability of these relationships to any material composition in conjunction with a simulation of wave propagation to establish coda wave interferometry as an early warning system for concrete structures.

Zusammenfassung

Die Zustandsüberwachung von Betonbauwerken ist aus Gründen der Sicherheit und Wirtschaftlichkeit eine gesamtgesellschaftliche Aufgabe, bei der zunehmend Techniken aus dem Bereich der zerstörungsfreien Prüfung zum Einsatz kommen. Eine vielversprechende Technik im Bereich der Dauerüberwachung von Strukturen aus heterogenen Materialien wie Beton oder Stahlbeton ist die Codawelleninterferometrie, da sie sensitiv gegenüber kleinsten Änderungen im Material ist und daher als Frühwarnsystem für auftretende Schäden in Frage kommt. Die Technik basiert auf dem Vergleich zweier Wellenformen im Ultraschallbereich, aus dem eine Geschwindigkeitsänderung abgeleitet werden kann. Diese ist aufgrund der großen Streuung in heterogenen Medien in spät eintreffenden Wellenpaketen, der sogenannten Coda, besonders gut messbar und auflösbar.

Ziel der Studie war es, die Beziehungen zwischen instationären Einflüssen, d. h. durch Temperatur, relative Luftfeuchtigkeit und externe Spannungsänderungen, auf die Geschwindigkeitsänderung von sich in Beton ausbreitenden Codawellen zu untersuchen. Zudem wurden die Langzeitphänomene Schwinden, Kriechen und die Hydratation mittels der Geschwindigkeitsänderung charakterisiert. Die in drei Artikeln veröffentlichten Untersuchungen zeigten, dass Temperatur- und Spannungsänderungen im elastischen Bereich linear mit der Geschwindigkeitsänderung der Coda zusammenhängen. Die Wechselwirkung von Temperatur und Feuchte hingegen führte zu einer nichtlinearen Änderung der Geschwindigkeit der Coda. Außerdem wurde gezeigt, dass die Prozesse Schwinden, Kriechen und Hydratation neben konventionellen Messtechniken auch mit der Codawelleninterferometrie beschreibbar sind.

Bei der Anwendung der Codawelleninterferometrie ist es entscheidend, die einzelnen Effekte voneinander trennen zu können, um eine reversible Geschwindigkeitsänderung von einem dauerhaften Schaden unterscheiden zu können. Für diese Trennung wird ein Ansatz angegeben, der die möglichen Einflüsse zusammenfasst. Alle entwickelten Zusammenhänge wurden unter Verwendung eines einheitlichen Betonrezepts entwickelt, sodass die Ergebnisse auf ein multiskaliges Modell übertragen werden können. Ein solches Modell kann im Zusammenspiel mit einer Simulation der Wellenausbreitung die Grundlage für die Übertragbarkeit dieser Zusammenhänge auf beliebige Materialkompositionen bilden, um die Codawelleninterferometrie als Frühwarnsystem für Strukturen aus Beton zu etablieren.

Contents

1 Introduction

1.1 The Big Picture

One of the greatest disasters in recent building history was the sudden collapse of the Polcevera Viaduct in Genoa, colloquially known as Ponte Morandi, which was an essential connection element between Northern Italy and France (Figure 1). The inner-city bridge was inaugurated in 1967 as a representative and innovative cable-stayed bridge comprising three pylons made from prestressed concrete, one of which collapsed on August 14, 2018 and caused the death of 43 people, followed by massive economical damage for the region [1].
Contrary to the preliminary assumptions of exceptionally unfavorable weather conditions on that day, several studies [2, 3, 4, 5] concluded that the main reason for the collapse was the combined effect of fatigue at a very high number of load cycles and corrosion that had been amplified by the saltiness from the sea and pollutants of nearby industry. The degradation of the bridge had advanced much more than expected at the time of its construction, and the over 80 million larger load amplitude cycles for the bridge were many more than originally assumed. Totally, over half a billion vehicles crossed the bridge during its service life [2].

Figure 1: Collapsed approximately 250 m long segment of the Polcevera Viaduct (Ponte Morandi) over the Val Polcevera connecting A10 four-lane motorway in Genoa, Italy. The particularly damaged pillar 9 of the western pylon triggered the structural failure causing 43 casualties on August 14, 2018. The remains of the original bridge were demolished in June 2019 and replaced by the Genoa-Saint George Bridge, inaugurated on August 3, 2020.

Furthermore, it was suggested that the collapse could have been predicted, and thus, avoided by adequate maintenance based on model predictions, valid at that time, or a structural health monitoring system using sensors that continuously tracked the condition of the bridge by means of non-destructive testing (NDT) methods [3, 5]. The authors of the later published studies specifically pointed out the generally underestimated phenomenon of combined fatigue and corrosion for structures built from reinforced and prestressed concrete to the scientific community.
The collapse also raised concerns about the general condition of infrastructure in Europe, and several countries demanded renovation and replacement of structures exposed to corrosion and deterioration.
On the basis of experience to date, only a tiny minority of structures made of reinforced concrete suffer from durability issues on the order of magnitude represented by the Polcevera Viaduct. The vast majority of our existing infrastrucutre is effectively monitored by regular standardized inspections, which depend on national bulding codes regarding their frequency and level of detail. Additionally, the knowledge about design principles to counter durability problems increased significantly, e.g., by the introduction of concepts for life cycle assessment [6, 7, 8, 9, 10], and consideration of data from installed monitoring systems for the evaluation of structural reliability [11, 12].
Using Germany as an example, the largest proportion of our infrastructural buildings was built using structural concrete, and the majority of the over 52,000 bridges were constructed between 1960 and 1990 [13]. Amongst novel reliability models and updated building codes, the scientific community addresses the safety needs of the ageing infrastructure, which has to deal with a steadily increasing traffic volume at the same time, by developing new monitoring methods for damage detection.
Structural concrete poses a special challenge to monitoring techniques, mainly because of its distinctive heterogeneity and complex microstructure. One of the most promising techniques for detecting even weak changes in the microstructure of concrete is ultrasound-based coda wave interferometry (CWI).
The time-dependent progress of corrosion-induced deterioration shown in Figure 2 [6] illustrates that microstructural changes, i.e., micro-scale damage formation, starts after depassivation of the reinforcement. The damage level becomes critical after exceeding a limit state for the crack width [14] because self-reinforcing effects occur due to facilitated gas and liquid transport mechanisms until the structure reaches the Ultimate Limit State, which is equal to its failure. Within this process, detection of mechanical damage is desired as early as possible for safety and economical reasons. Detection is accomplished by either inspection at discrete points in time or by a continuous monitoring system.

Due to its extraordinary sensitivity in heterogeneous media like concrete, CWI is more suitable for damage detection at an unprecedented early stage than conventional NDT techniques.

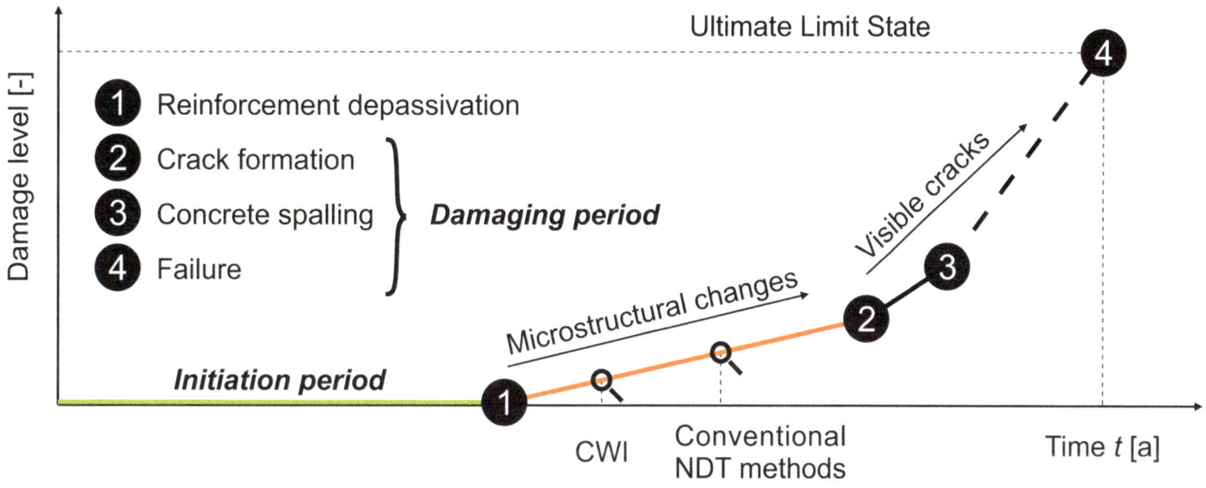

Figure 2: Time-dependent progress of corrosion-induced deterioration and damaging of reinforced concrete structures and their limit state (adapted from [6]), characterized by the initiation period and the damaging period, during which mechanical damage occurs. In comparison with many conventional non-destructive testing methods in heterogeneous media, coda wave interferometry is more sensitive and has the potential to detect damage before other techniques.

1.2 Research Objective

The potential sensitivity of coda waves to weak changes in heterogeneous media has been recognized long before its application to concrete structures. Early studies from the geophysical community investigated the seismic source, attenuation and scattering effects of coda waves based on small local earthquakes as early as 1969 [15, 16]. Finite difference and finite element methods have already been suggested for the case of known model parameters to interpret seismograms with respect to geometrical localization of any given change. The model parameters for the Earth's crust were typically unknown, however, they may be well-known in the current static models used for engineering structures. Later in 2002, CWI was introduced by revealing the relationship between the seismic velocity in granite and temperature changes, together with the associated acoustic emissions [17]. Furthermore, the coda waves, which initially appear to be chaotic and noisy, were found to be highly repeatable as no medium change appears [18].

The exceptional sensitivity, together with the signal repeatability, led to a large number of studies regarding the detection and localization of weak changes in concrete using CWI with respect to ultrasonic wave speed, as affected by thermal, mechanical and chemical damage [19]. However, these impacts may affect a structure at the same time and add various terms to the total velocity change. This superimposition is a big challenge regarding the interpretation of the velocity variation as their exact separation among the individual effects is unknown. Irreversible changes are of particular interest because they are generally synonymous with occuring damage in the structure. Therefore, another aspect of interpreting the velocity variation is the separation between changes that are reversible, as well as those which are irreversible and indicate damage. Figure 3 shows different impacts on engineering structures that affect the properties of coda waves. The general research objective in this study is the characterization of the transient impacts of temperature, moisture, and stress on coda waves, as expressed by the ultrasonic velocity change dv/v.

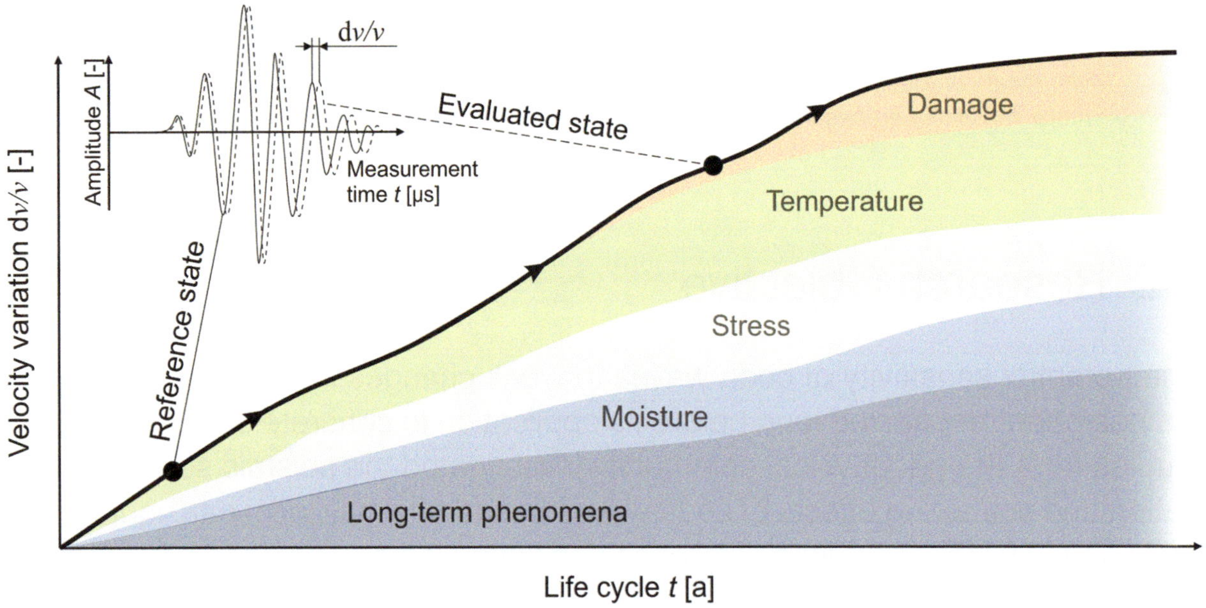

Figure 3: Summarized effects on the velocity variation dv/v between two ultrasonic signals propagating in a structure at a reference state and at the evaluated state. The term long-term phenomena aggregates further concrete-specific phenomena, e.g., shrinkage, creep, carbonation, and corrosion, whereas the remaining effects act on a short-term scale.

Several advances are expected due to investigating the relationships between the various impacts on the velocity variation in a systematic way, i.e., using a consistent concrete mix and identical raw materials throughout all of the experiments. In addition to increasing the level of knowledge about CWI data analysis and potential measurement setups for both laboratory and in-situ studies, this study also covers the following three major theses:

- We gain an improved understanding of the physical relationships regarding a state change in concrete and an ultrasonic velocity change of the propagating coda waves by means of controlled experiments on a laboratory scale, during which we only vary one impact parameter at a time. We can quantify the developed relationships by the change in dv/v per impact using the stretching method [20].

- We can transfer the developed relationships to the structural scale to translate between a change in the measured total dv/v and a state change.

- We create input data for material models on the micro- and mesostructural scale that are capable of simulating nonlinear state changes, including damage. These models are the basis for wave propagation simulations used to attribute changes in dv/v to occurring damage. The relationships developed will be transferable to arbitrary concrete mixes and systems by means of these models.

The research presented in this study addresses the formulated hypotheses by way of three articles, which are summarized in the final conclusions. The results will contribute to establishing the CWI method with regard to applicability as an early-warning monitoring system for integrity assessment of structures made of reinforced concrete. The motivation for the presented research is the need for a general safety increase in our built environment.

1.3 Research Framework

This study originated within the framework of the research unit *CoDA – Concrete Damage Assessment by Coda Waves* [21], which was founded in 2019. The research unit is organized as a mainly national network of scientists coming from the disciplines of geophysics and civil engineering, and funded by the *German Research Foundation*. Its overall objective is the development of a novel methodology for continuous condition assessment of reinforced concrete structures, which is capable of continuously updating durability-oriented computational models [22].
Research within this unit is conducted on both the specimen scale and the structural scale. The specimen scale focuses on exploring the effects of microstructural changes on the coda using experiments in a laboratory environment, which is covered within this study. Simultaneously developed material models [23, 24] form the basis for high-performance simulations of wave propagation [25, 26, 27] to represent the experimentally observed changes. Furthermore, experiments

on real structures on the structural scale [28, 29, 30, 31] were conducted with the objective of identifying and localizing any occurring material changes using CWI-based techniques [32, 33, 34, 35] combined with numerical structural models.

2 Theoretical Background

2.1 Concrete Raw Materials and their Elastic Properties

The design process of concrete has many degrees of freedom. Possible considerations reach from the planned area of application, which primarily implies requirements regarding mechanical strength and exposition, over rheological properties of the fresh concrete, and the on-site construction method. Technical considerations can be further extended with respect to the environmental impact, or the related emissions of the materials, and their availability, as well as economical aspects. At the end of the design process, concrete is defined by a mix of several raw materials in the desired ratio. The basic components are cement, acting as a binder, plus mixing water, and aggregates of various sizes. Besides the main components, admixtures, and additives can be added, or substitute other components [36].
Cement is a ground, inorganic hydraulic binder that hardens during the chemical reaction of water with several phases, mainly calcium silicate. This reaction is referred to as hydration. Depending on the ratio of water and cement w/c, non-hydrated cement particles either remain in the microstructure for a ratio below approximately 0.40, while excess water, which remains in the system above a ratio of 0.40, facilitates the formation of capillary pores because it evaporates slowly over time [37]. Different cement compositions can be classified according to building standards, e.g., *DIN EN 197-1* [38] or *ASTM C150-07* [39], depending on the ratio of the main components, which are Portland clincer, blast-furnace slag, pozzolan, fly ash, burnt shale, limestone, or silica.
Cement paste is the hardened product of cementitious materials that have reacted with water. Its compressive strength linearly depends on the degree of hydration above a critical hydration degree [40]. However, strength increases rather slowly above this threshold, and the macroscopic properties are considered to be a combination of the properties of the individual cement phases. An excerpt of the properties of various characteristic minerals is provided in Table 1 [41, 42]. Information regarding the elastic properties of further phases is summarized in the literature [43]. In addition to hardened cement paste, elasticity and yield stress during the fresh state of concrete are becoming increasingly important properties due to their industrial use, e.g., for 3D printing technologies, of extremely fluid cement-based materials such as self-compacting or self-leveling mortars, and concretes [44].

Table 1: Excerpt of elastic properties of characteristic cement and cement paste phases with their cement chemist notation (CCN) [43, 41, 42]: compression (bulk) modulus K [GPa], shear modulus G [GPa], Young's modulus E [GPa], and Poisson ration ν [-].

Cement phase	CCN	K	G	E	ν
Tricalcium silicate	C_3S	105.2	44.8	117.6	0.134
Dicalcium silicate	C_2S	105.2	44.8	117.6	0.134
Tricalcium aluminate	C_3A	105.2	44.8	117.6	0.134
Silica fume	SiO_2	36.5	31.2	72.8	0.167
Portlandite	CH	40.0	16.0	42.3	0.324
C-S-H	$C_{1.7}SH_4$	14.9	9.0	22.4	0.25

The last basic component of concrete are aggregates, where conventionally used crushed or naturally mined stones still make up the largest share of the aggregates in concrete. They can be characterized by several physical properties, e.g., their pore space, density, magnetic properties, radioactivity, thermal, and electrical properties.
The elastic and seismic properties are the basis to describe wave propagation in isotropic solid materials [45], e.g., by means of the link of the ultrasonic velocities, and mechanical stresses provided by the acousto-elasticity theory [46]. Generally, the wave velocity v follows Equation (2.1) using the quotient of the wave path s over time t.

$$v = \frac{s}{t} \tag{2.1}$$

The ultrasonic pulse velocity in concrete is generally frequency dependent [47], whereby frequency affects both wave speed and attenuation [48]. Equation (2.2) describes the frequency f using the wavelength λ.

$$f = \frac{v}{\lambda} \tag{2.2}$$

The elastic wave velocities of the basic types of body waves, the primary wave (pressure wave or P-wave) with velocity v_p and the secondary wave (shear wave or S-wave) with velocity v_s are given in Equations (2.3a) and (2.3b) using the material density ρ, Lamé's first parameter λ, and Lamés second parameter μ.

$$v_p = \sqrt{\frac{\lambda + 2\mu}{\rho}} \tag{2.3a}$$

$$v_s = \sqrt{\frac{\mu}{\rho}} \tag{2.3b}$$

Concrete can be represented by material models, which consist of a system of cement paste, aggregate particles, the pore system, and increasingly a solid phase that is considered as the perturbed region around the aggregate particles. This Interfacial Transition Zone (ITZ) incorporates slightly different transportation mechanisms, and mechanical characteristics [49, 50, 51].
To directly relate microstructural material models, which use locally discretized material parameters, with elastic properties, a homogenized stiffness tensor, e.g., of the hydrated cement paste, can be determined. A formulation of the tensor can be accomplished by means of effective medium theories, e.g., using a Mori-Tanaka scheme [52]. These properties are, in turn, connected with elastic wave velocities using the effective bulk modulus K and the shear modulus G, which is equal to μ [53].

$$K = \rho \left({v_p}^2 - \frac{4}{3} {v_s}^2 \right) \tag{2.4}$$

$$G = \rho {v_s}^2 \tag{2.5}$$

In addition, the dynamic modulus of elasticity E_d can be determined non-destructively by means of the wave velocities [54], where the dynamic Poisson's ratio ν_d can be estimated around 0.26 for low standard and 0.18 for high strength self-compacting concretes [55].
However, the dynamic modulus of elasticity is around 30 % higher than the conventionally determined static modulus [56] because E_d is determined at low stresses and is therefore rather a tangent modulus.

$$v_p = \sqrt{\frac{E_d}{\rho} \frac{1-\mu}{(1+\mu)(1-2\mu)}} \tag{2.6a}$$

$$v_s = \sqrt{\frac{E_d}{\rho} \frac{1}{2(1+\mu)}} = \sqrt{\frac{G}{\rho}} \tag{2.6b}$$

The elastic properties of naturally occurring aggregates are subject to variations. They are governed by the properties of their individual minerals, their fractional volume, and their bonding, temperature, and pressure. Therefore, the properties can be anisotropic and deviate from macroscopic moduli, assuming homogeneous stress and strain [57, 58]. Table 2 provides elastic properties for characteristic rock-forming trigonal and Feldspars minerals for aggregates in concrete. Additionally, wave velocities and Poisson's ratio of recycled aggregates are being increasingly considered due to their reusability [59].

Table 2: Elastic properties of characteristic rock-forming minerals [60, 61], calculated according to Reuss (index R) [58] and Voigt (index V) [57]: density ρ [kg/m^3], mean atomic mass m_A [amu], compression (bulk) modulus K [GPa], Young's modulus E [GPa], shear modulus G [GPa], velocities of the P- and S-waves v_p [m/s] and v_s [m/s], and Poisson ration ν [-].

System	ρ	K_R	E_R	G_R	v_p	ν
Mineral	m_A	K_V	E_V	G_V	v_s	
trigonal						
α-quartz	2650	38.0	91.8	41.5	6050	0.08
	20.03	38.6	101.2	47.6	4110	
Calcite	2712	71.4	68.8	25.7	6660	0.31
	20.02	77.7	92.6	35.6	3390	
Feldspars						
Albite	2605	53.0	69.0	26.9	6060	0.28
	20.17	61.6	80.5	31.4	3350	
Anorthite	2760	79	-	37	7050	0.29
	-	89	-	43	3800	
Orthoclase	2570	40.3	55.8	22.0	5690	-
	-	56.0	78.3	30.9	3260	

The velocity of commonly used magmatic and metamorphic rocks has a range that depends on the individual mineral composition. For typical aggregates in concrete, compression and shear wave velocities range between $v_p = 2.46...5.56$ km/s and $v_s = 2.07...3.49$ km/s for Quartzite ($\rho = 2636$ kg/m^3) and $v_p = 4.67...6.15$ km/s and $v_s = 2.61...3.51$ km/s for dry Basalt ($\rho = 2840$ kg/m^3).

In addition to the main components, finely ground admixtures are often added either as inert (type I), e.g., ground aggregates or pigments, or reactive materials during the hydration (type II), e.g., fly ash, blast-furnace slag, or silica powder [36]. Admixtures of type II can partially substitute the volume of necessary cement. Therefore, these supplementary cementitious materials (SCM) are increasingly being used to create more eco-efficient cements [49, 62]. For example, current research aims at replacing an increasing portion of Portland cement clinker by agricultural waste products like risk husk ash [63, 64].

In order to influence the concrete properties, additives can be added to the concrete mix. Currently, almost every state-of-the-art concrete used in practice is being optimized regarding its compactability, workability, and durability [36].

The group of additives comprises effects on the fresh concrete that are either water-reducing, retarding, accelerating, or have further effects according to *ASTM C494/C494M-17* [65]. Eventually, the addition of these additives results in adjusted properties of the hardened concrete, e.g., compressive or flexural strength, elastic modulus, or durability-related properties, e.g., resistance to freeze-thaw cycles.
Besides its advantageous properties regarding mechanical strength, reinforced concrete (RC) structures can be subject to physical and chemical attacks, which have an adverse impact on their durability and therefore cause a degradation of the material. RC durability is predominantly governed by its resistance against phenomena on a microstructural level, such as freeze-thaw attack [66, 67, 68], carbonation [69, 70, 71], or chloride penetration along with moisture transport and the resulting reinforcement corrosion [6, 72, 73]. All of these phenomena cause deterioration and crack formation, which can amplify the entire damaging process besides crack development due to internal stresses [74] [75]. To the best of our knowledge, this process must be controlled in addition to the overall condition of the structure for safety, and reliability reasons starting with a proper concrete design.

2.2 Wave Propagation Theory in Heterogeneous Solids

A mechanical wave is characterized by oscillations of individual particles coupled to each other and the associated acoustic energy transport through a medium. Figure 4 (a) shows a sinusoidal oscillation of a single particle over time, represented by a mass connected to a spring that moves around a neutral position due to an initial deflection. The relationship between the force F acting on the particle and the deflection x is given by Hooke's law by means of a constant k for the spring.

$$F = -kx \tag{2.7}$$

The connection of an arbitrarily large set of masses to each other creates a simplified model of a homogeneous, elastic solid with Young's modulus E, where each particle is able to oscillate and transfer energy to the adjacent particles (Figure 4 (b)). In this model, a phase shift exists between the particles reached earlier and the ones that are reached with lag time. Assuming that the particle motion and the wave propagation have the identical direction, both compaction and dilution zones appear, and the resulting wave type is called a pressure wave. If the particle motion is, however, perpendicular to the direction of propagation,

the result is a shear wave. These are the basic two types of waves that can appear in a solid body without any boundaries or interfaces.
Besides pure body waves, surface waves such as Rayleigh or Love waves form a group of waves, which propagate along an interface between differing media and diminish for increasing distance from the surface. Typically, they propagate slower than body waves but can have larger amplitudes [76].

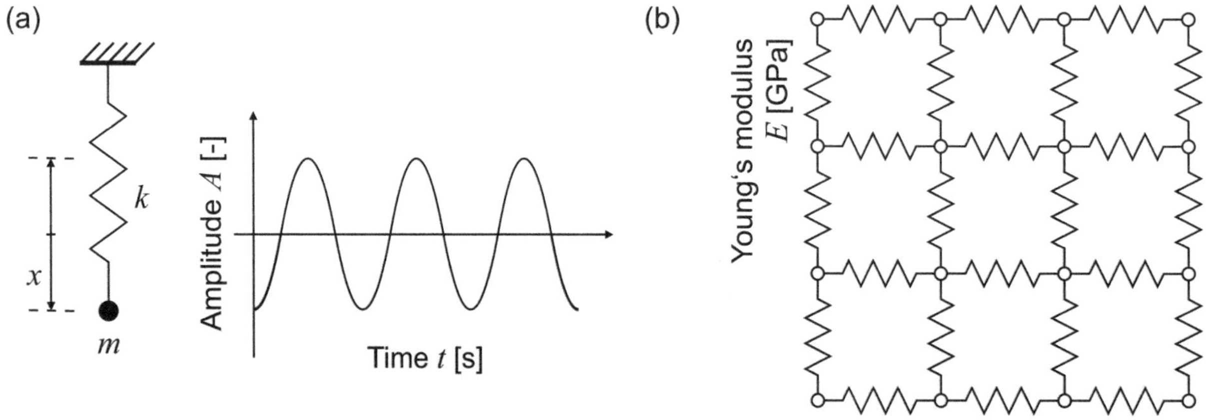

Figure 4: (a) Sinusoidal oscillation of a linear spring according to Hooke's law with an initial deflection without attenuation effects. (b) Simpflified model of an elastic, homogeneous, and isotropic solid with particles connected by linear springs. In the model, the spring constant is equal to Young's modulus E [76].

Using Hooke's law and Newton's second law of inertia, the equation for wave propagation in elastic solids can be derived [77]. This equation describes the propagation of a sound wave in elastic media through the displacement of a particle.

$$\rho\frac{\partial^2\vec{u}}{\partial t^2} = (\lambda + 2\mu)\,\nabla\,(\nabla\cdot\vec{u}) - \mu\nabla\times\nabla\times\vec{u} \tag{2.8}$$

In addition to theoretical solutions of Equation (2.8), e.g., for the P-wave in Equation (2.3a) and the S-wave in Equation (2.3b), numerical methods such as discrete element method or finite difference-based methods exist, which allow approximated solutions even for solids with a high degree of heterogeneity and geometrical boundaries. The basis for such simulations are realistic morphologies of the material, which reflect the elasticity and the moment of inertia for the elements of a discretized mesh. Figure 5 (a) shows a computationally generated mesostructure given an aggregate, pore, and crack distribution for a concrete cuboid [24, 23], e.g., using the concrete mesostructure generator *pyCMG* [78].
Figure 5 (b) [25] shows a simulation using the discrete element method of a complex ultrasonic wavefield between two embedded transducers in such a mesostructure.

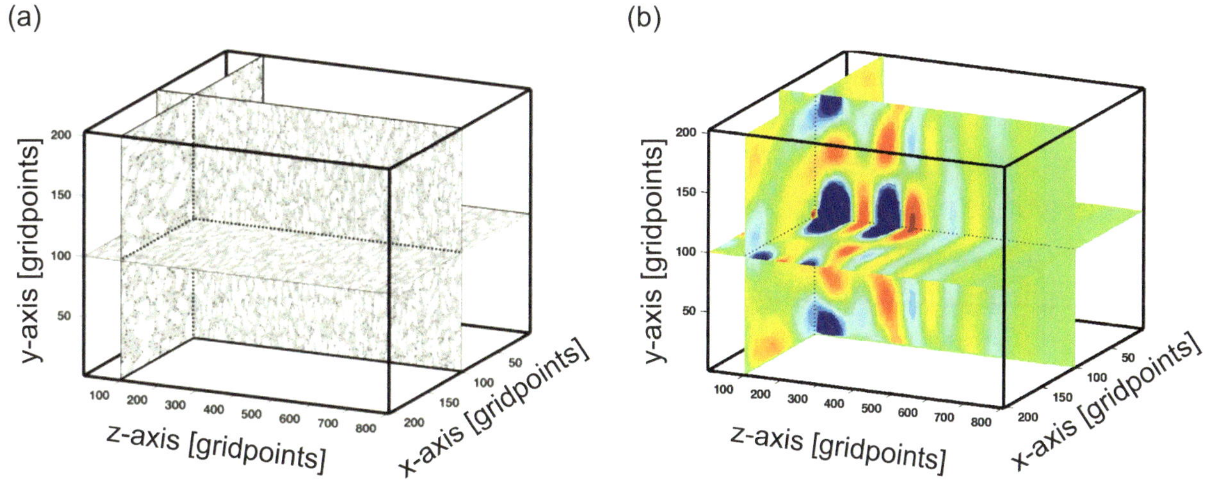

Figure 5: (a) Computationally generated concrete model for mesoscale and multiscale simulations containing aggregates, cement paste, and air voids [23]. (b) Simulation of a complex wavefield between two transducers in a concrete model for damage identification using the discrete element method [25].

Sound waves in heterogeneous media can be excited either naturally, e.g., by crack formation in concrete, or by means of an artifical energy source. Assuming a point source, the energy spreads spherically from the source, and its intensity I decreases proportional to r^2 according to spherical divergence:

$$I \propto \frac{1}{r^2} \tag{2.9}$$

Since the intensity is proportional to the square of the amplitude A of the wave, the amplitude decays with the radius:

$$A \propto \frac{1}{r} \tag{2.10}$$

For the fully elastic model in Figure 4 (b), elastic waves can propagate without energy loss, and the amplitude decays only due to spherical divergence. However, heterogeneous media such as concrete exhibit attenuation effects, which amplify the amplitude decay over distance traveled. Attenuation is governed by the distinct physical phenomena intrinsic absorption and scattering. Intrinsic attenuation describes the conversion of the wave's mechanical energy into heat due to friction and changes in viscosity. Scattering effects occur at interfaces, at which the wave is split into reflected and transmitted wave modes. This effect intensifies for a large number of scatterers, creating long and complicated wave paths on which energy is converted as a result of absorption [77]. Altogether, the amplitude decay of elastic waves propagating in concrete can be described using a diffusion approach [79] or a radiative transfer approach [80].

Scattering in concrete increases particularly fast if the size of the scatterers, e.g., aggregates, pores, or cracks, is approximately as long as one-tenth of the wavelength [76]. Many scattering events consequently lead to complex and diffuse wavefields, as seen in Figure 5 (b). Since conventional techniques, based on the evaluation of unperturbed waveforms, often fail when used in heterogeneous media, a need for the development of backscattering techniques such as coda wave interferometry arises, which involves and simulaneously exploits the multiple scattering effects.
Attenuation also increases considerably with frequency, so that the reasonable range for elastic wave-based inspection, e.g., ultrasonic testing, is limited to a range of frequencies [81]. Four different regimes can be distinguished, which are described in greater detail in the following paragraphs [19]. Table 3 shows the frequency domains for concrete, including the assumed paths of an elastic wave, the regime name, the approximate range of relevant frequencies dependent on the particular concrete mix, and the dimensions of the wave and the structure.
The stationary wave regime is characterized by low frequencies below 20 kHz, in which the wavelength is greater or around equal to the size of the structure. Techniques operating in this regime, such as modal analysis, use the vibrational eigenmodes of the evaluated structure. These techniques are usually robust but less sensitive compared to techniques using higher frequencies. They can detect changes only by means of differing eigenmodes without explicitely resolving the material change due to the relatively large wavelengths.
The simple scattering regime covers waves that interact weakly with the concrete heterogeneities with frequencies between 10 kHz and 150 kHz. This range is considered to be the standard range for most ultrasonic testing techniques, in which the wavelength is smaller than the structure but still larger than the vast majority of the scatterers. The frequency range in the simple scattering regime aims at finding the balance between reasonable sensitivity to material changes on the one hand, and a wide volume coverage in the structure with a high signal-to-noise ratio in the waveform due to low attenuation on the other hand.
In the multiple scattering regime, elastic waves interact strongly with the concrete heterogeneities at frequencies higher than 100 kHz, whereby the transition between the simple and the multiple scattering regimes and their respective frequency limits are smooth. The mesoscopic frequency range above around 100 kHz creates waveforms that can be described by random scattering. For such waveforms, the evaluation of the backscattered tail of the ultrasonic signal, the coda, becomes increasingly informative due to information reinforcement over a large number of wave paths.

Table 3: The four frequency domains of ultrasound in concrete [19] with wave paths between a sender S and a receiver R. The stated frequency limits have large uncertainties because they depend on the geometrical dimensions of scatterers in concrete, e.g., aggregates, and air voids. The ultrasonic wavelength λ is compared to the specimen length L, the size of the heterogeneities d, the scattering mean free path l,and the intrinsic absorption characteristic length l_a.

Wave paths	*Regime*	*Frequency*	*Dimensions*
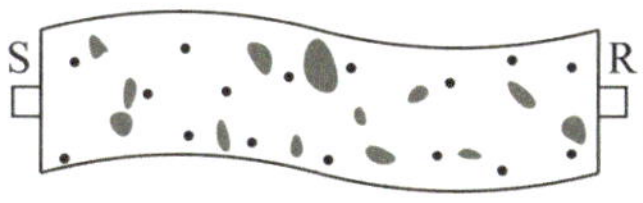	Stationary wave regime	< 20 kHz	$\lambda > L$
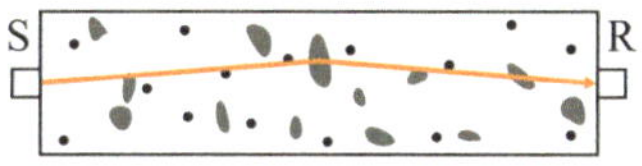	Simple scattering regime	10 kHz-150 kHz	$d < \lambda < L < l_a$
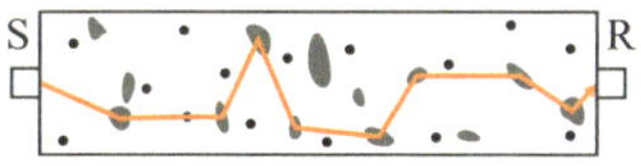	Multiple scattering regime	100 kHz-1 MHz	$\lambda < d \leq l < L < l_a$
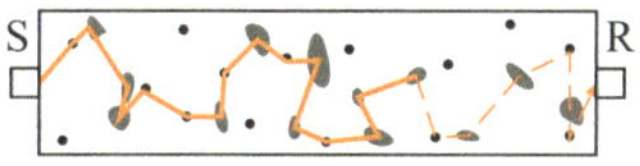	Attenuation regime	>1 MHz	$\lambda \ll L$

The attenuation regime is characterized by frequencies above 1 MHz, for which elastic waves are strongly attenuated, and the wavelength is much smaller compared to the size of the structure. Consequently, the distances traveled by the waves become very small, and the structure size scales down to experiments in the laboratory.

2.3 Coda Wave Interferometry: State of the Art

2.3.1 Theory of Coda Wave Interferometry

Typical concrete structures include a large number of scatterers, in which NDT techniques based on direct or simple scattering assumptions fail. Therefore, frequencies in the multiple scattering regime can be selected using backscattering techniques such as coda wave interferometry for estimating nonlinear behavior in seismic velocity [18].

Coda waves refer to the late arriving tail of a waveform, which is strongly scat-

tered, and characterized by both a high degree of diffusivity and high sensitivity at the same time. Originally studied in the geophysical community as a result of small local earthquakes [15, 16, 82], coda waves were also proven to be sensitive to even weak material changes using ultrasound experiments [17].
In contrast to their chaotic and noisy appearance, e.g., in complicated waveforms, they were found to be highly reproducible and deterministic. Therefore, the medium in which coda waves propagate is often referred to as a seismic microscope, or interferometer, in which small changes between two states can be detected. The theory of coda wave interferometry is comprehensively summarized in the literature and is the basis for the following paragraphs [18].
The fundamental assumption of the theory of coda wave interferometry is path summation. Every given wavefield $u(t)$ can be expressed by the sum of waves along all wave paths traveled $S_P(t)$ [83]. For any interpretation of the wavefield, this assumption implies that the individual wave paths do not have to be known exactly with respect to further considerations.

$$u(t) = \sum_P S_P(t) \tag{2.11}$$

Individual waveforms can be compared with each other, e.g., by choosing a late time window of the waveform, in which multiple scattering can be expected. Any perturbation between two waveforms consequently indicates a perturbation in the medium. As long as the scattering properties in the medium do not change, the perturbation can be assigned to displacements of the scatterers, e.g., by a change of the path length δl or the introduction of new scatterers.
The perturbed wavefield is given by an adaption of Equation (2.11) by a change in travel time τ_P, induced by a phase change. This theory holds true as long as the scattering mean free path is considerably larger than a single wavelength [18].

$$\tilde{u}(t) = \sum_P S_P(t - \tau_P) \tag{2.12}$$

The magnitude of the change in travel time depends on the scattering path P, whereby the chosen time window for the waveform limits the evaluated paths. The perturbed waveform can be compared to a reference waveform, as in Figure 6, by means of the time-shifted correlation coefficient $CC(t_s)$ using a finite time window between t_1 and t_2 [18]. Here, t_s is the time shift between the compared waveforms.

$$CC(t_s) = \frac{\int_{t_1}^{t_2} u(t)\tilde{u}(t+t_s)\,dt}{\sqrt{\int_{t_1}^{t_2} u^2(t)\,dt \int_{t_1}^{t_2} \tilde{u}^2(t+t_s)\,dt}} \tag{2.13}$$

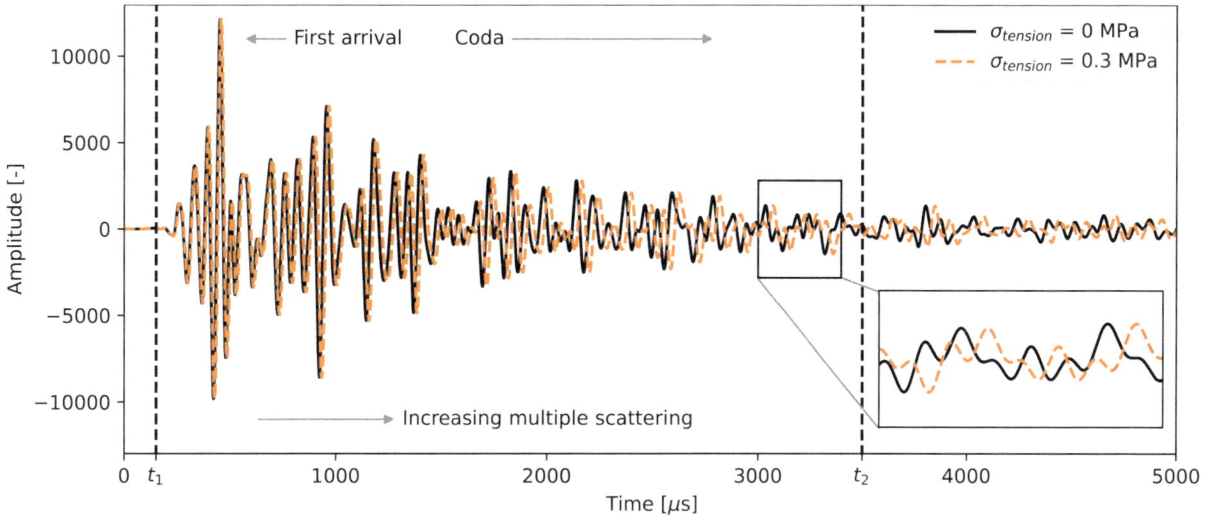

Figure 6: Ultrasonic signal in concrete under tension compared to a reference signal at an unloaded state, evaluated within the time window between t_1 and t_2. The load induces a perturbation in travel time that accumulates over time. The change can be resolved and determined increasingly well for late arrivals in the coda.

Equation (2.13) was developed considering doublet techniques that are able to measure temporal variations of crustal velocities based on comparing pairs of microearthquakes using a moving time window and computing differences in arrival times [84, 85]. However, greater stability is reached for velocity-based evaluation of any perturbation. Using the wave velocity v in a medium and the unperturbed travel time t_P, the change in travel time τ_P can be related to a relative velocity perturbation dv.

$$t_P + \tau_P = \int_P \frac{1}{v + dv} ds \tag{2.14}$$

The state of the art of comparing two ultrasonic waveforms in concrete is the stretching technique [86] using the velocity change dv/v. The evaluated waveform is stretched in time by a factor $\varepsilon = -dv/v$, while simultaneously computing a correlation between the stretched and the reference waveform. For a maximum correlation, the stretching factor can be indicated:

$$CC(\varepsilon) = \frac{\int_{t_1}^{t_2} u(t)\tilde{u}(t(1-\varepsilon))\,dt}{\sqrt{\int_{t_1}^{t_2} u^2(t)\,dt \int_{t_1}^{t_2} \tilde{u}^2(t(1-\varepsilon))\,dt}} \tag{2.15}$$

Finally, it should be noted that the travel time is, in fact, the average arrival time of the waves that arrive in a finite time window, and t_P can therefore be replaced with t. Furthermore, the inferred velocity change is also the weighted average of velocity changes in both elastic wave types, the P-wave, and S-wave [17],

whereby the sensitivity of the coda is generally governed by changes in the S-wave velocity [16].
Furthermore, any perturbation between two waveforms not only induces a phase change but also wave significant waveform distortion due to attenuation effects, i.e., a change of the waveform. The quantification of this effect using the quality factor Q is further described, and discussed in Chapter 3.2.2.

2.3.2 Effects of State Changes on the Coda

Coda wave interferometry was the subject of numerous studies in the area of concrete monitoring since its sensitivity even to weak changes was proven [17, 18]. Various effects on the velocity change of coda waves were characterized such as temperature [87, 88] or mechanical loads [89, 90, 91, 92, 93], and even typical concrete phenomena, e.g., freeze-thaw attack [94], self-healing [95], or alkali silica reaction [96]. The effect of moisture changes in concrete on the velocity variation using CWI is hardly studied, probably because of the large amount of time required to systematically collect experimental data due to slow transport processes in concrete. However, water content affects ultrasound propagation in concrete [97, 98] and must therefore be considered when comparing ultrasonic waveforms by means of CWI, at least for signals that are far apart in time.
Although single impacts on the velocity variation can be characterized, the separation of the individual effects is hardly discussed but becomes indispensable when distinguishing the sources of the perturbation in the investigated medium, i.e., the state change. For on-site application of CWI, this is becoming increasingly important because structures are exposed to several influencing variables at the same time.
Nevertheless, approaches exist to determine the effect of various simultaneously acting variables on concrete specimens in the laboratory. Using the example of external mechanical stress, the acoustoelastic effect [46, 99, 100] provides the relationship between the applied uniaxial stress in z-direction σ_{zz} compared to a reference state with wave velocity v_{ij}^0, and velocity in the medium v_{ij} using acousto-elastic constants A_{ij} in the elastic regime.

$$v_{ij} = v_{ij}^0(1 + A_{ij}\sigma_{zz}) \tag{2.16}$$

Figure 7 illustrates the relationship between the velocity change and a stress change when experimental data is available and a regression can be performed.

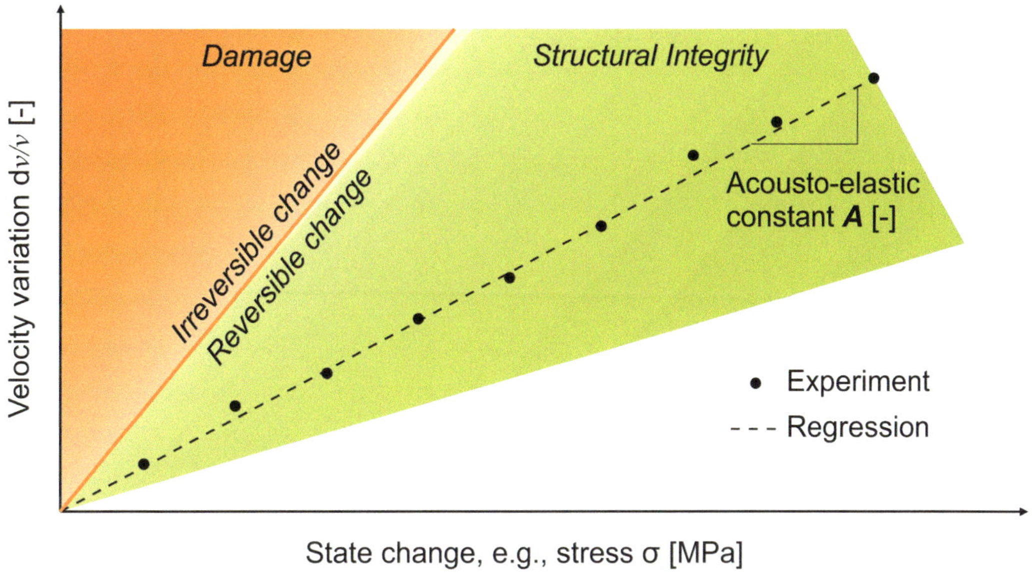

*Figure 7: Relative velocity change for a state change, adapted from [19]. The relationship, for which structural integrity is assured, can be obtained by a regression of experimentally determined data. For an external uniaxial load on a concrete structure, the velocity change is linearly coupled to the load by the acoustoelastic constant **A** stating a linear relationship.*

Calibration using a regression analysis only becomes possible as soon as experiments can be conducted and further considerable interactions are excluded in the data. In this case, any values for the velocity change that deviate too much from the calculated regression values can be considered as an indicator for damage. However, the location of the boundary to which extent deviations from the regression are considered to be still acceptable is difficult to indicate and currently not precisely specified.

Any additional variable affecting the structure such as temperature would have to be compensated to apply Equation (2.16) because it only depends on the applied stress. Therefore, a bias control technique was proposed to compensate for additional effects and the bias originating from experimental procedures on the velocity variation using a reference specimen for each additional effect [101, 102]. This approach is very well suited for structures on the laboratory scale, but both the collection of experimental data, and the compensation using a reference specimen with identical behavior regarding wave velocity changes can be challenging in practice. The study presented here aims at determining experiment-based relationships, using a consistent concrete mix to explore the impacts on the velocity change and to prepare a coherent dataset for modeling and wave propagation simulations in arbitrary concrete mixes in order to separate the individual effects from each other.

3 Impact of Temperature and Relative Humidity Variations on Coda Waves in Concrete [1]

Abstract - *The microstructure of concrete can be affected by many factors, from non-destructive environmental factors through to destructive damage induced by transient stresses. Coda wave interferometry is a technique that is sensitive enough to detect weak changes within concrete by evaluating the ultrasonic signal perturbation compared to a reference state. As concrete microstructure is sensitive to many factors, it is important to separate their contributions to the observables. In this study, we characterize the relationships between the concrete elastic and anelastic properties, and temperature and relative humidity. We confirm previous theoretical studies that found a linear relationship between temperature changes and seismic velocity variations for a given concrete mix, and provide scaling factors per Kelvin for multiple settings. We also confirm an anti-correlation with relative humidity using long-term conditioning. Furthermore, we explore beyond the existing studies to establish the relationship linking humidity and temperature changes to seismic attenuation. Lastly, we formulate a model for the combined impact of temperature and relative humidity on the signal velocity variation.*

---

[1] This article was published in the peer-reviewed journal *Scientific Reports* by *Nature Research* on 14 August 2024. This chapter is the originally submitted version, which was accepted by the Technical University of Munich. The authors' work share is listed in Appendix 2.

3.1 Introduction

Today, large parts of the built environment are made from structural concrete. The safety of these structures is, in most cases, secured during regular, but often only annual, inspections rather than using continuous monitoring. Thus, these maintenance strategies rely on information from an ideally full update of the structures' integrity on discrete inspection dates. In contrast, monitoring techniques have the advantage of continuously detecting structural changes and, therefore, allow preventive action to be initiated at an early stage before critical damage occurs. One recent and promising technique for monitoring heterogeneous media such as concrete is Coda Wave Interferometry (CWI) [84, 17, 18]. CWI leverages the fact that scattered waves propagate multiple times in perturbed media, summing the perturbations and changing the arrival times of waves relative to a reference state.

Multiple laboratory-scale studies have shown the technique's potential to monitor microstructural changes in concrete as a consequence of external mechanical loads [103, 104, 105, 106] or due to temperature changes [107, 87]. Furthermore, time-dependent concrete-specific phenomena can be characterized using CWI, e.g., the alkali-silica reaction [108] or freeze-thaw damage [94]. Naturally, some of these investigations have been extended to large-scale concrete structures [109, 110, 34, 35, 31], where CWI's potential has been proven under controlled environmental conditions.

Besides its high sensitivity to even weak changes in the material, one of the biggest challenges for CWI today is the simultaneous occurrence of condition changes that cause superimposed terms contributing to the velocity variation. The separation of these terms is often difficult under practical conditions because changes in environmental conditions can occur simultaneously (e.g., winters are wetter and colder in the mid-latitudes). This applies especially to interpreting the effect of transient changes, e.g., due to stresses or damage, because environmental temperature or humidity changes may induce a bias in measuring the ultrasonic signals' phase lag [111, 112, 113, 114].

In on-site applications, one specific challenge arises from the time-limited evaluation period (often only hours) because this is required so that only one parameter is changing [19]. However, to link the impact of temperature and damage, nonlinear ultrasonic techniques have been suggested either to link the influence of damage on ultrasonic waves using thermal modulation [115] or to compensate for temperature effects using a thermal bias control by comparing the temperature-induced perturbation of test and reference specimens [101, 102].

In our study, we address the necessity of separating the superimposed effects of environmental conditions affecting every concrete structure, i.e., temperature

and relative humidity changes, on the speed at which seismic waves propagate. We detected the changes in the coda waves using long-term equilibrium experiments and remained in the non-destructive regime to avoid having to address any additional impact of damage on the velocity variation. We separated the individual effects of temperature and relative humidity from each other and proposed a third-order polynomial regression model for their combined effects to predict velocity variations.
The specimens are a normal concrete mix, which is described in detail in [24, 23]. In addition to elastic wavespeed perturbations, monitoring the scattering properties of the materials, by measurements of amplitude decay in the coda waves, may be useful because the amplitude decay is particularly sensitive to damage. Therefore, we also demonstrate the changes in anelastic seismic properties (attenuation and scattering) in the seismic quality factor Q [15, 16] for both temperature and relative humidity variations.

3.2 Materials and Methods

3.2.1 Coda Wave Interferometry

Coda wave interferometry is based on comparing deterministic and reproducible ultrasonic waveforms propagating in a heterogeneous medium along multiply scattered paths [19]. Any change in the medium's scattering properties causes a perturbation of the waveform compared to the original, unperturbed waveform. Even weak changes can accumulate over many repeated wave paths between a sender-receiver pair and can be detected in late coda waves through phase differences, which can be expressed as a velocity perturbation dv/v of the ultrasonic signal, i.e., the velocity variation. The velocity variation is deduced from a linear regression of the time delay, which increases over signal time t, and the measurements of phase differences, dt. It is, therefore, most significant when measured in late windows that record scattered waves called the ultrasonic coda.
Multiple methods for CWI analysis have recently been proposed [17, 116, 117, 118], based on deriving a stretching factor $\varepsilon = dt/t = -dv/v$ from the calculation of a maximum signal correlation coefficient CC between a reference signal $u(t)$ and a perturbed signal $\tilde{u}(t)$. Here, t_1 and t_2 mark the time window in which the velocity change is derived. In this study, we use the stretching technique [116] to determine relative velocity changes dv/v in the medium according to Equation (3.1).

$$CC(\varepsilon) = \frac{\int_{t_1}^{t_2} u(t)\tilde{u}(t(1-\varepsilon))\,dt}{\sqrt{\int_{t_1}^{t_2} u^2(t)\,dt \int_{t_1}^{t_2} \tilde{u}^2(t(1-\varepsilon))\,dt}} \tag{3.1}$$

The ultrasonic signals were preprocessed and were effectively noise-free. We refer to [119] for details regarding the preprocessing procedure, where the value of $CC(\varepsilon)$ generally depends on the signal-to-noise ratio and the evaluated time-window. The cross-correlation coefficient and velocity variation were computed across the entire duration of each experiment using a window length of 2,500 samples, equal to 2.5 ms.

3.2.2 Quality Factor

The quality factor Q describes the fractional loss of energy of an ultrasonic signal arriving at a receiver position. As such, it is a parameter representing the attenuation properties of a medium in a single backscattering model and was originally studied to evaluate the attenuation of coda waves from small local earthquakes [15, 16]. It is dimensionless, frequency-dependent, and can be expressed as in Equation (3.2), which is valid for both the single scattering and diffusion theories. The amplitude of an ultrasonic signal $A(\omega|t)$ at a frequency around ω over a time t is expressed by a source term c, which is constant for a uniform excitation. The constant a accounts for geometrical spreading, where $a = 1$ is generally assumed for body waves.

$$A(\omega|t) = ct^{-a}e^{-\omega t/2Q} \tag{3.2}$$

As we use embedded sensors, we can assume that the ultrasonic coda waves are mainly energy transported by body waves. We can rewrite the equation by replacing ω with the frequency f, around which the signal is filtered.

$$A(f|t) = ct^{-1}e^{-\pi ft/Q} \tag{3.3}$$

By transforming the equation, the right side of Equation (3.4) can be interpreted as a linear function with a negative slope of $\pi f/Q$.

$$\ln\left[A(f|t)t\right] = \ln c - \frac{\pi f}{Q}t \tag{3.4}$$

The procedure for deriving Q is illustrated in Figure 8 for an ultrasonic signal, where the root mean square of the amplitude is the basis for linear regression, starting from the maximum amplitude of the filtered signal.

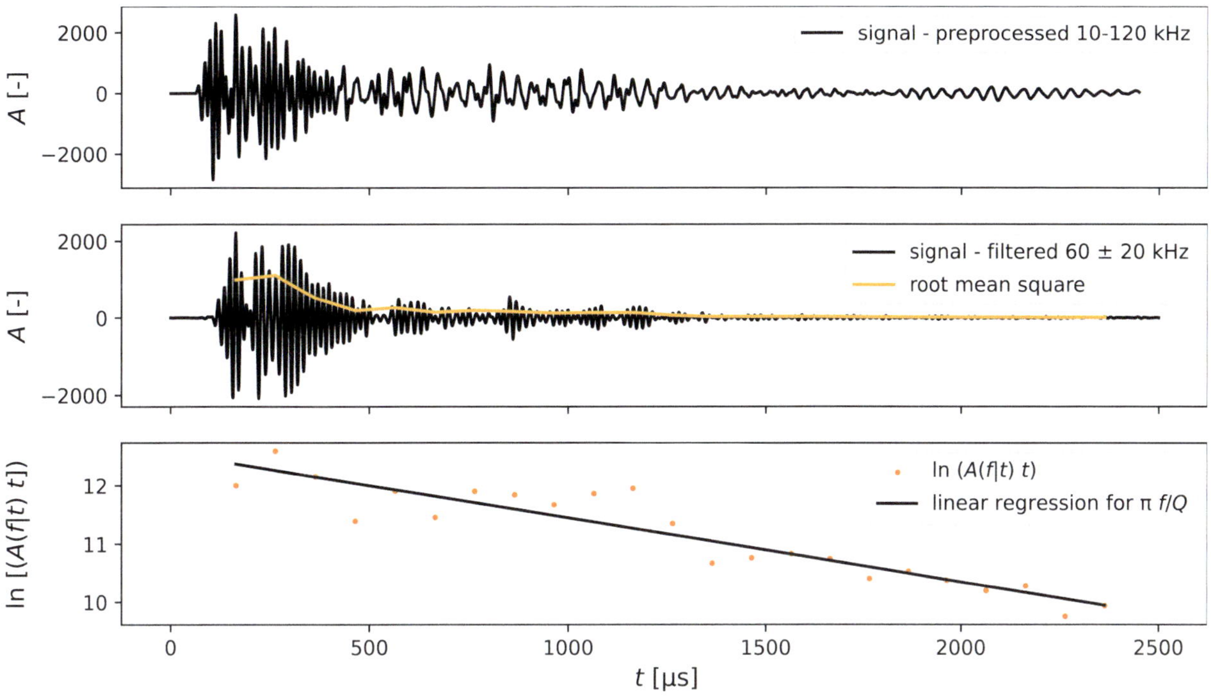

Figure 8: Procedure for deriving the quality factor Q using the signal's root mean square [120]. Starting from the preprocessed signal, the signal is further filtered around a central frequency f, e.g., 60 kHz with a bandwidth of $\pm$20 kHz. An approximation for Q according to Equation (3.4) *is the slope of the linear regression using the root mean square of the signal amplitude A for a constant f.*

The computation of Q according to Equation (3.2) is affected by reflections at geometrical boundaries that can be observed in the form of periodic wave packages in the root mean square in Figure 8. Q is lapse-time dependent [121], which we do not explore here. Therefore, our absolute estimate of Q is rather uncertain and we focus on the relative measurements.

3.2.3 Materials

We produced two batches of concrete specimens with constant cement content and aggregate size distributions (Table 4). Only the water-cement ratio was varied, inducing an increased pore space volume within the cement paste matrix for greater values of w/c. The cement used in both mixes was Ordinary Portland Cement *CEM I 42.5 R* and the aggregates were primarily quartzitic and crushed. To decrease the viscosity of the fresh concrete and enhance the coupling of the embedded sensors, we added 2.4 % of polycarboxylatether-based superplasticizer to the concrete with $w/c = 0.45$. A standardized pressure method test yielded an air void content of 2.0 % air voids for $w/c = 0.45$ and 0.8 % air voids for $w/c = 0.60$ in the fresh concrete.

After production, the specimens were covered with water-retaining jute mats for 24 h and further cured under water for more than 100 d. The levels of chemical and autogenous shrinkage were minimal due to the high degree of hydration [122, 123] and thus negligible for the subsequent experiment. However, drying shrinkage along with carbonation shrinkage [124, 125] are significant while decreasing the relative humidity as part of the experimental setup and are discussed later in the article.

Table 4: Raw materials and their composition for the two concrete mixes with $w/c = 0.45$ and $w/c = 0.60$, where the aggregates are primarily quartzitic and crushed. The cement type is CEM I 42.5 R and the superplasticizer is BASF MasterEase 3880.

Cement Content	**Water / Cement**	**Aggregate Size**				**Superplasticizer**
CEM I 42.5 R	w/c	0/2 mm	2/5 mm	5/8 mm	8/16 mm	BASF ME 3880
350 kg/m³	0.45	40 %	12 %	29 %	19 % *	2.4 % **
350 kg/m³	0.60	40 %	12 %	29 %	19 % *	0

* Contains 1.2 % oversized grains 16.0/22.4 mm.
** As a mass fraction in % of the cement content.

3.2.4 Experimental Setup

To investigate the impact of both temperature and relative humidity on the ultrasonic coda, we varied the ambient parameters temperature T_{amb} and relative humidity RH_{amb} while monitoring the concrete specimens' internal condition using embedded sensors and mass tracking. The set of ambient and embedded sensors, together with the geometrical setup, is presented in Figure 9 (a). It consisted of ambient temperature and relative humidity sensors in the climate chamber and an embedded Pt100 element for temperature T [°C] measurement positioned at the center of each specimen. Strain gauges (*PL-60-11-3LJCT-F*) with a gauge resistance of $R = 120\,\Omega$ and a length of $l = 60\,mm$, applied to the surface on two opposite sides and implemented as a quarter bridge, tracked the strain. In addition, the mass m of specimens from the same batch was continuously tracked. Three specimens were monitored for each T, m, and the ultrasonic signals.

The ultrasonic signals were retrieved from symmetrically positioned embedded transducers, each at a distance of 300 mm from the center axis (Figure 9 (a)), which act as sender-receiver pairs with a center frequency of 62 kHz [91], proposed for CWI in concrete structures [126]. The measurement system [127] continuously samples signals with a length of 12,000 samples at a rate of 1 MHz and a 14 bit resolution.

Figure 9 (b) shows the experimental setup of concrete specimens with embedded sensors in the climate chamber. Additional specimens without any measurement equipment are regularly weighed to track changes in mass, i.e., water loss. Each specimen set, for the ultrasonic measurements, for the temperature and strain measurements, and for the mass tracking contained three specimens.

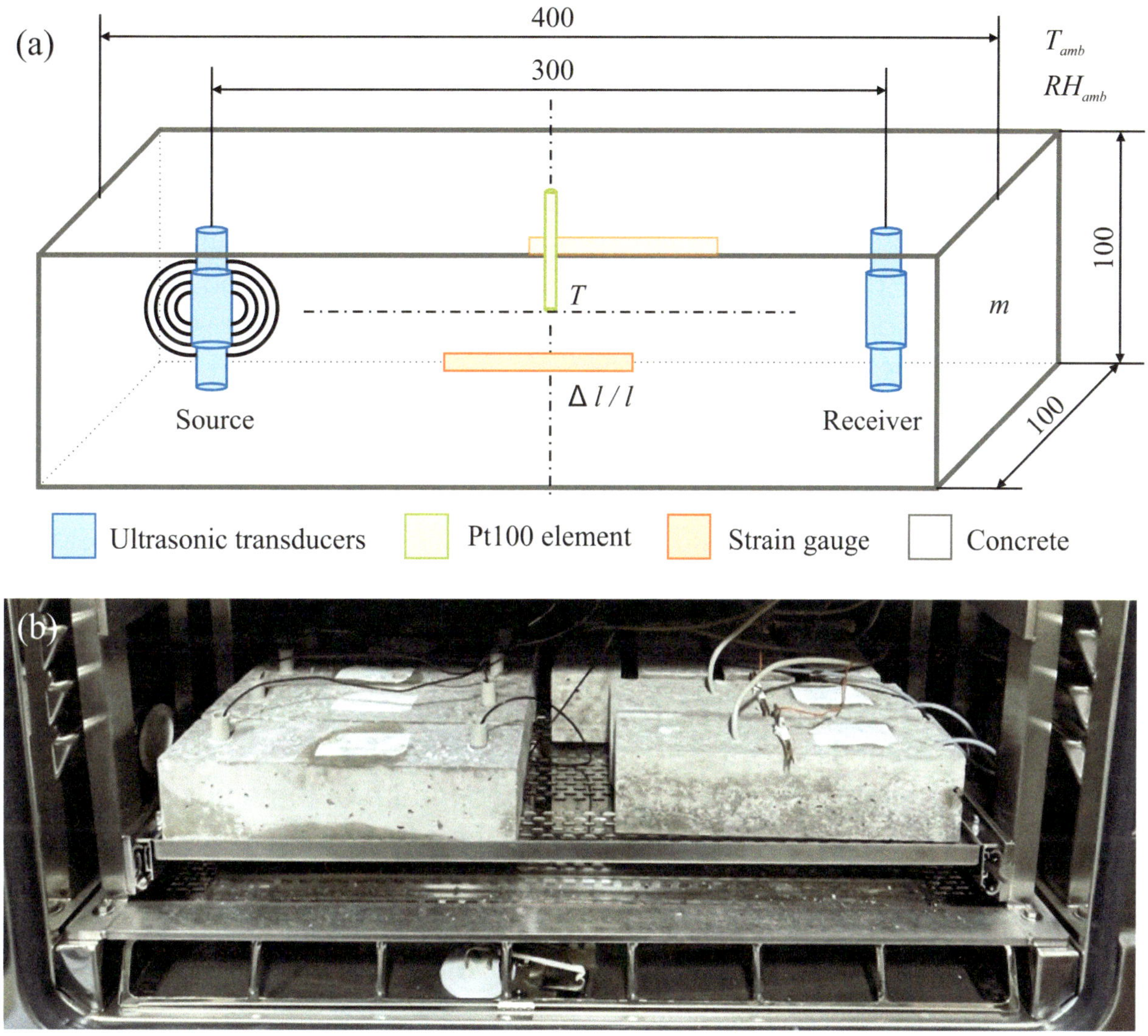

Figure 9: (a) The geometry of the specimens for all conducted investigations is a cuboid with an edge length of 400 mm and a square cross-section of 100 x 100 mm² For both concrete mixes ($w/c = [0.45, 0.60]$), three specimens are equipped with ultrasonic transducers and another three specimens with a Pt100 element (T). The specimens' mass m is monitored for three further specimens. (b) Photograph of the experimental setup in the climate chamber at controlled ambient temperature T_{amb} and relative humidity RH_{amb}.

The experimental variations in temperature and relative humidity had to be chosen carefully because several mechanisms other than the non-destructive boundaries trigger damage in the concrete matrix, thus contributing to superimposed changes in the ultrasonic velocity.

Several processes can introduce damage on a microstructural level due to thermal loading [128]. These can be physicochemical processes, i.e., the decomposition of hydrates [129], by microchemically induced cracking in the interfacial transition zone between the aggregates and the cement paste due to differential dilation [130], by drying shrinkage due to differential strains [131], by increasing vapor pressure [132] or by their gradients [133]. When elevating the temperature, ettringite is the first hydrate phase to decompose. It is a crystalline thermally unstable component of Portland cement, and the decomposition of its phase-pure form is discussed in the literature for combinations of temperature and vapor pressure [134].

We varied the temperature between 55 °C and 2 °C, while the relative humidity was modulated between 95 % RH and 35 % RH avoiding any ettringite decomposition. Further decomposition mechanisms beyond the stated temperatures and relative humidity do not apply because the study's objective is to determine the relations in the non-destructive regime.

Figure 10 presents the modulation of the external control variables T_{amb} and RH_{amb} in the climate chamber together with the specimens' mass changes Δm, where the starting conditions are 95 % RH at 20 °C. The relative humidity is held constant at a temperature of 20 °C until the mass change decreases below a threshold of 0.05 % per week. After reaching a nearly constant mass, we assumed a constant water saturation in the concrete matrix and introduced a stepwise temperature variation starting at 55 °C. We decreased the temperature in 5 °C steps with a last step of 3 °C down to a minimum of 2 °C. Each temperature was held for a 5 h duration to condition the entire specimen at the same temperature, validated by equalization of the temperature of the embedded central sensor with the ambient temperature. After the temperature variation, the relative humidity was decreased by 15 % until a final step that reached 35 % RH. The monitored strain matched the predicted strain from the model according to *DIN EN 1992-1-1* [135]. We conclude that drying shrinkage is responsible for the measured strain in our experiment.

3.3 Results

3.3.1 Temperature Impact on the Coda

Ultrasonic measurements were conducted as the temperature was varied between 55 °C and 2 °C once the specimens had reached a constant relative humidity state, which was validated by their steady mass. The results for each relative humidity state are plotted in Figure 11 for two concrete mixes with dif-

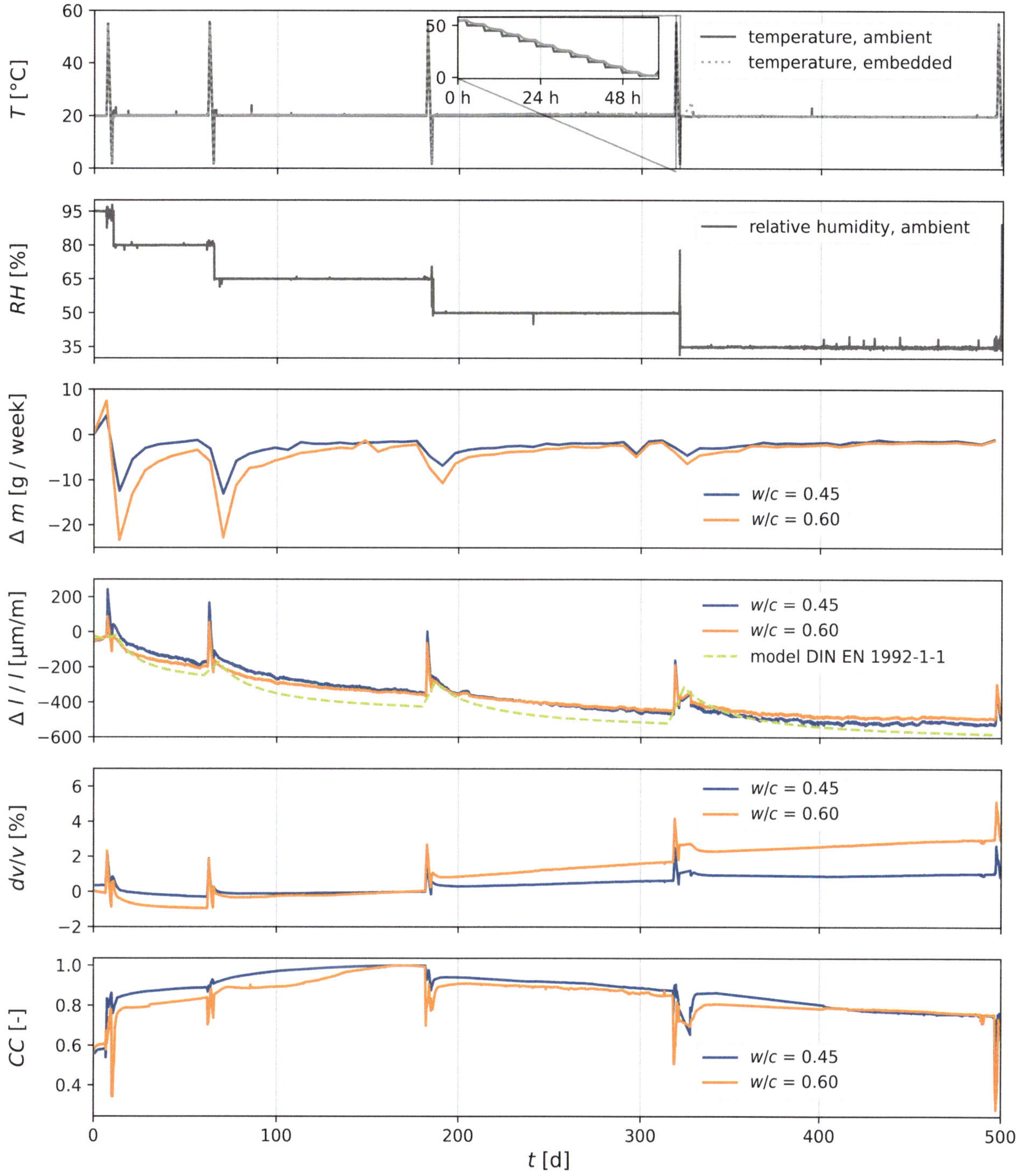

Figure 10: Experimental measurements with the ambient and embedded control variables T and RH. The mass losses Δm for both specimen series $w/c = 0.45$ and $w/c = 0.60$ were monitored weekly, and daily immediately after each temperature variation cycle. Strains $\Delta l/l$ were monitored throughout the experiment based on the mean of readings taken from two strain gauges (quarter bridges) and compared to a model for drying shrinkage according to DIN EN 1992-1-1 *[135]. Throughout the experiment, the velocity variations dv/v were computed based on Equation* (3.1).

ferent w/c ratios. Their standard deviations, σ, and linear regressions are also shown. The slopes of the regression are indicated as the velocity variation per Kelvin, while the reference signal, used to normalize the velocity variation values, was selected at 20 °C for each temperature variation cycle.

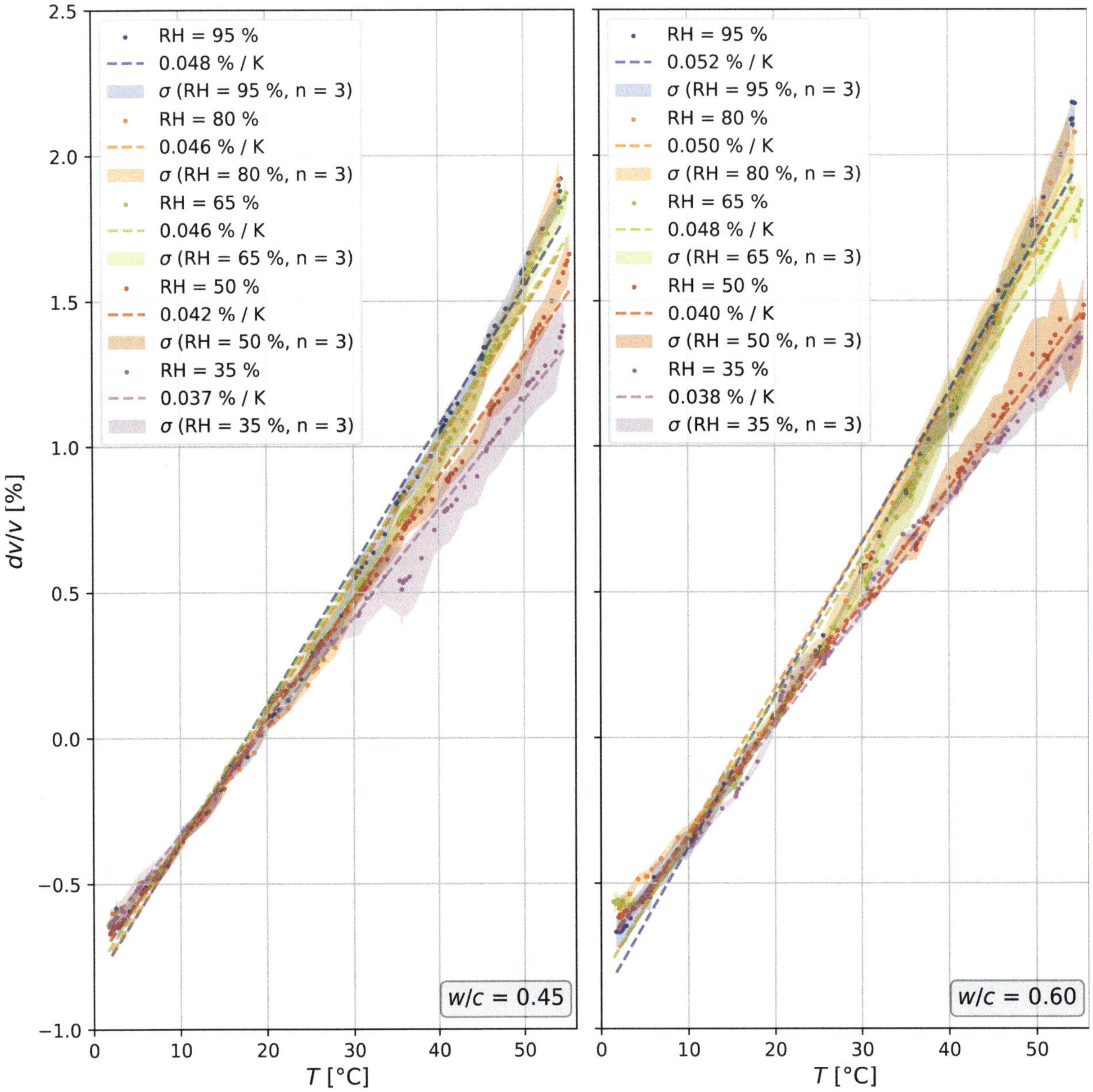

Figure 11: Velocity variation dv/v of the ultrasonic signals plotted against temperature for the two concrete mixes with $w/c = 0.45$ and $w/c = 0.60$. Dots represent the data points, whereas the dashed line is a linear regression for each data set. The R^2 values are all greater than 0.985 for all mixes and states. The transparent fill shows the standard deviation σ for all three specimens. The reference signals are selected approximately at a temperature T of approximately $20\,°$C.

For both investigated concrete mixes, the slope of the velocity variation plots steadily decreases as the relative humidity decreases, because the degree of saturation is lower. However, we note a larger gap between maximum and minimum slopes for the mix with $w/c = 0.60$. The difference between the two mixes is particularly notable at higher saturation levels. As the mixes only differ in the ratio w/c, we can derive an interdependency between the change in the slope and the concrete's pore system. The pore volume, as well as the portion of the water stored in it, and the average pore size are larger for $w/c = 0.60$. For this concrete, more water evacuates from the pore system in between the temperature cycles due to more dynamic water transport processes, thus resulting in an increase in the slope.

3.3.2 Relative Humidity Impact on the Coda

Only constant-state ultrasonic signals, validated by a nearly constant mass, were evaluated to derive the relationship between the velocity variations and the relative humidity of the specimens. We intentionally ignored the transition times between humidity regimes in order to ignore fluid diffusion processes and focus on steady-state regimes. Figure 12 shows, for both concrete mixes, the correlation between dv/v and RH at a constant temperature $T = 20\,°\text{C}$ for ten signals immediately before each temperature variation cycle as depicted in Figure 10.

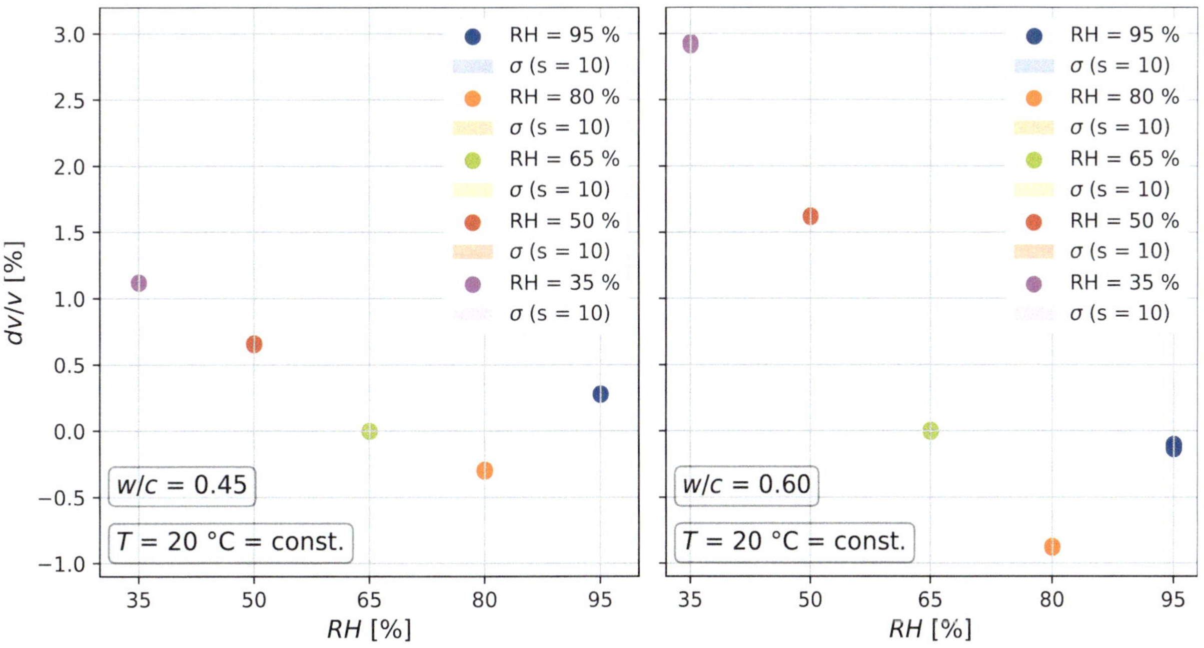

Figure 12: Velocity variation dv/v of the ultrasonic signals plotted against relative humidity for the two concrete mixes with $w/c = 0.45$ and $w/c = 0.60$. Dots represent the data points, where the standard deviations for the ten signals collected at half-hourly intervals are smaller than the dots in this figure.

For both mixes, the reference signals, used to normalize the velocity variation values, were chosen at 65 % RH. Figure 12 shows similar trends for both mixes: dv/v decreases with increasing concrete relative humidity until a value of 80 % is reached. The value at 95 % RH is an exception for both mixes because in this state, almost all pores are filled and dv/v is higher than at 80 % RH. Once again, the mix with $w/c = 0.60$ exhibits larger differences between the RH states, which we interpret as a consequence of a higher pore volume and more extensive transport processes.

3.3.3 Combined Impact of Temperature and Relative Humidity

The results from Figure 11 and Figure 12 with variations of only one impact variable form the basis for investigations on the combined effect of temperature and relative humidity on the coda for a variable w/c ratio. Data from both sets were matched for $T = 20\,°C$, where we defined a boundary condition in Equation (3.5).

$$dv/v\,(T = 20\,°C\,|\,RH = 65\,\%) = 0 \tag{3.5}$$

Using polynomial regression, the measurement data were fitted to the following third-order polynomial over three variables, the water-to-cement ratio w/c, the temperature T, and relative humidity RH (Equation (3.6)). This fitting procedure uses the data for compensation when analyzing ultrasonic measurements and evaluating them under varying ambient conditions.

$$dv/v = \sum_{i=0}^{i=3}\sum_{j=0}^{j=3}\sum_{k=0}^{k=3} a_{ijk}\left(\frac{w}{c}\right)^{i} T^{j} RH^{k} \tag{3.6}$$

In the above equation, a_{ijk} refers to the polynomial coefficients that were fitted to the measurement data. The corresponding values are provided in Table 5 in the appendix. The predictions of the model for values of the temperature, relative humidity, and water-to-cement ratios for the whole range of values within which the measurement data is available are shown in Figure 13. We provide the metrics of the model fit, where the R^2 value is 0.999, and the mean-squared error is 1.5604×10^{-7}. The above polynomial equation was obtained using the ordinary least squares method by performing linear regression on the polynomial features. The sklearn python package was used for performing the regression. To identify the most sensitive components of the polynomial, the dataset was standardized using the mean μ and the standard deviation σ in Equation (3.7).

$$x' = \frac{x - \mu(x)}{\sigma\,(x)} \tag{3.7}$$

The following expression (Equation (3.8)) for the relative velocity change ignores terms that are one order of magnitude smaller than the largest coefficient. The relative velocity change is most sensitive to the relative humidity, then the temperature, and finally, the water-to-cement ratio.

$$\left(\frac{dv}{v}\right)' \propto 0.26T' - 0.50RH' - 0.36RH'\left(\frac{w}{c}\right)' + (0.26T' - 0.49RH')\left(\frac{w}{c}\right)'^2 + 0.14RH'^3 \quad (3.8)$$

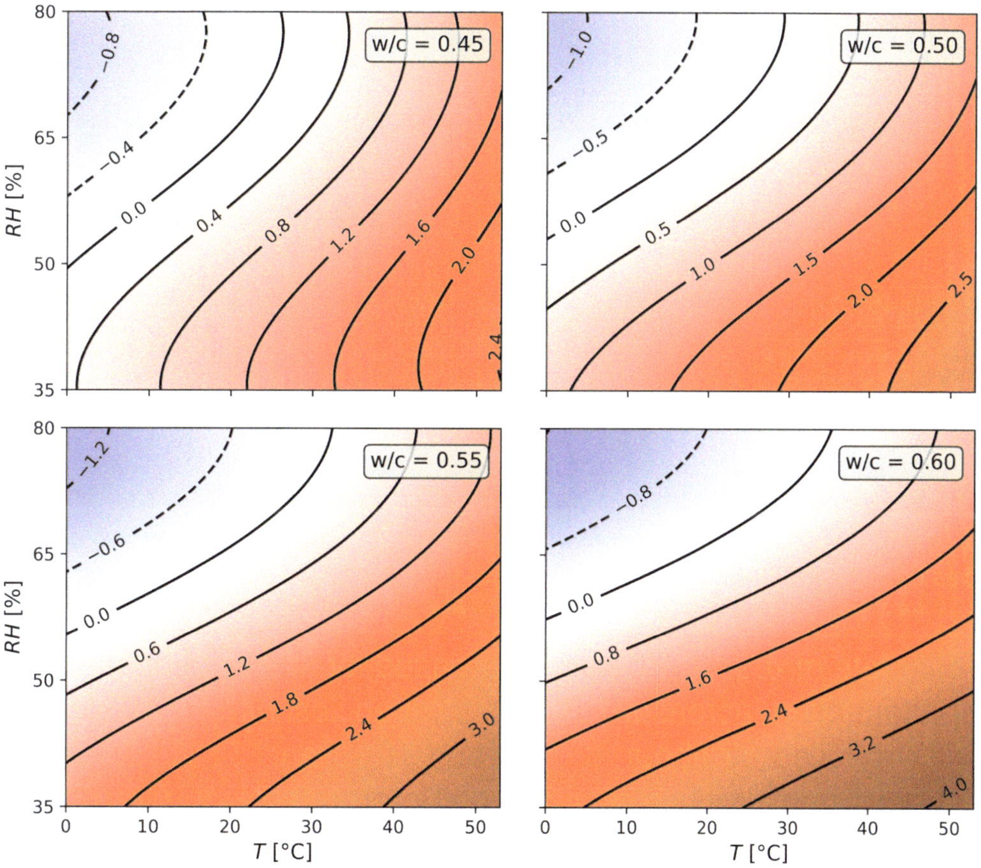

Figure 13: Contour plots of the velocity variation dv/v for four water-to-cement ratios dependent on temperature and relative humidity. The plots for $w/c = 0.45$ and $w/c = 0.60$ are based on measurements, while the plots for $w/c = 0.50$ and $w/c = 0.55$ are predictions of the third-order polynomial model. The lines in the plots are iso-lines of constant relative velocity variation.

3.3.4 Impact on the Quality Factor

The quality factor Q was computed based on the procedure in Figure 8 using Equation (3.2). First, Q was evaluated as a function of T for three specimens at two central frequencies (Figure 14): 60 kHz, which is approximately the resonance frequency of the sensor, and 90 kHz, which is still considered to be in the simple scattering regime of concrete [126]. Both signal filters had a bandwidth of 20 kHz around the central frequency.

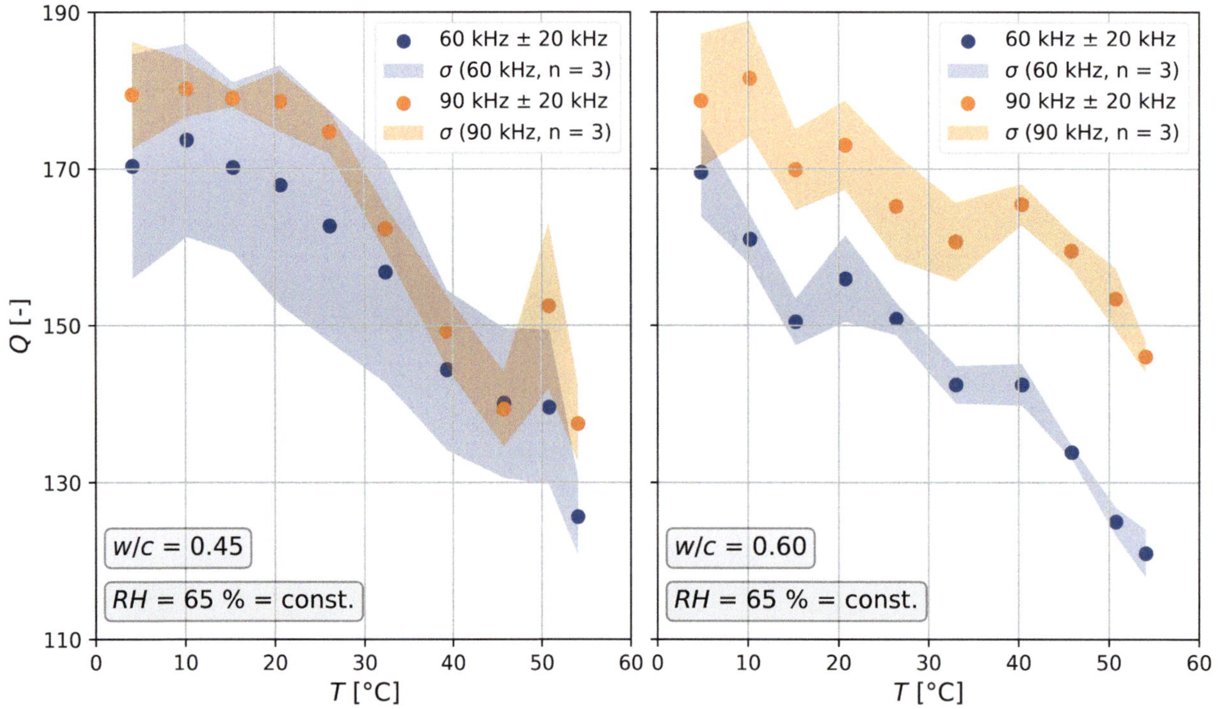

Figure 14: Quality factor Q for temperature variations for the two concrete mixes with $w/c = 0.45$ and $w/c = 0.60$, each preconditioned at $RH = 65\,\%$. Q was evaluated for two central frequencies 60 kHz and 90 kHz.

For both concrete mixes, Q generally decreased for higher temperatures. Thus, the signal attenuation increased for both frequencies with increasing temperature. Q was reduced by about 40 between 10 °C and 50 °C for both compositions. If the absolute values of Q are not interpretable due to multiple wave reflections in the specimen, the interpretation of the changes in Q should still be reliable. Additionally, we obtain a similar, continuous, decreasing trend in Q for increasing relative humidity for both frequencies and both mixes. The more saturated the pore space, the higher is the attenuation of the ultrasonic signals (Figure 15). Q has a significantly higher spread in values than when the temperature is varied: values of Q differ around 115 % over the measured humidity range. Therefore, the sensitivity of attenuation to relative humidity is much stronger than the temperature sensitivity.

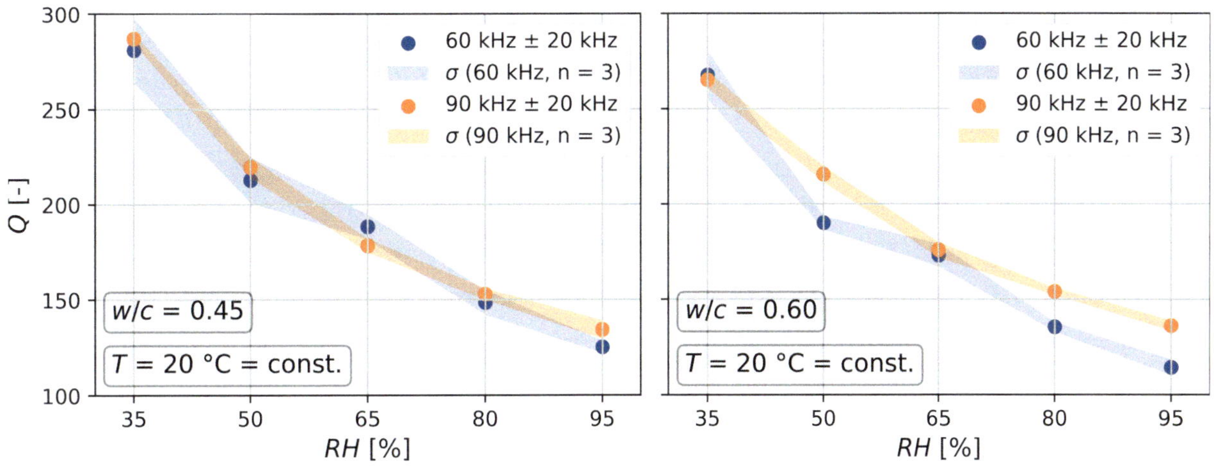

Figure 15: Quality factor Q for relative humidity variations for the two concrete mixes with $w/c = 0.45$ and $w/c = 0.60$ at a constant temperature of 20°C. Q was evaluated for two center frequencies 60 kHz and 90 kHz.

To show the sensitivity of the attenuation properties to T,RH, and w/c, we provide a standardized polynomial expression using the computed quality factors in the identical parameter space as used in Equation (3.8). For both increasing temperature and relative humidity, we identify a negative correlation with Q, thus, increasing attenuation.

$$Q \propto \left(-0.18T' - 0.38RH'\right)\left(1 + \left(\frac{w}{c}\right)^{\prime 2}\right) + 0.14T \cdot RH' - 0.11T'^2 + 0.18RH'^2 \qquad (3.9)$$

Despite the dependence in Q with lag time, frequency, and window length, this relation is the first to suggest a coupling of the temperature and humidity effects in affecting attenuation. Recent work from [136] also shows the negative correlation between Q and T and water content in sandstone. Their correlations exhibit strong dependence on the pore space of the sample, which is suggested and formulated by our regression.

3.4 Discussion

We evaluated non-destructive and reversible state changes in concrete using CWI, a technique characterized by high precision and sensitivity. Further, we evaluated the monitoring of anelastic properties in ultrasonic attenuation. We explored these states by changing the most commonly varying environmental factors, namely temperature and relative humidity. We found that temperature profoundly influenced ultrasonic speed and relative humidity on attenuation.
Where changes in ultrasonic velocity dv/v due to temperature are concerned,

our results are comparable with previous studies to within an order of magnitude [107, 137], although the values for the regression slopes (Figure 11) are slightly smaller in our study. The relationship between temperature and dv/v is quasi-linear for the same water-to-cement ratio, and simple linear regressions all show high R^2-values, all above 0.985, confirming an accurate fit to the experimental data. In our experimental design, we ensured a homogeneous temperature distribution over the entire volume of the specimens with embedded sensors and therefore avoided the influence of gradients in the material. Thanks to the relative humidity control of the climate chamber during the temperature variation and the relatively short duration of the temperature cycles, slightly more than two days, we minimized the effect of drying at the geometrical edges, which otherwise amplifies a hysteresis between cooling and heating cycles [137]. The impact of near-edge drying on dv/v is difficult to quantify and decreases for larger concrete sample geometries and greater distances between the structural edges and the sensors, where at least the first factor is generally the case for real civil engineering structures.

We also evaluated the impact of relative humidity variations on the material. We increased the experiment time significantly due to the slow diffusion transport processes in the concrete's pore system. We validated a relative humidity distribution that was sufficiently constant to meet our requirements by measuring changes in the mass of specimens, which is a precise and reliable measurement method. Nevertheless, we assume minor relative humidity gradients in the material decrease from the core. These gradients, however, still exist after years of conditioning, and we consider the introduced mass change threshold as a valid criterion for a nearly constant relative humidity distribution.

In addition, we compared the theoretical drying shrinkage (negative volumetric strains) based on *DIN EN1992-1-1* with our strain measurements in Figure 10. It could be demonstrated that the measured strain was approximately equal to the predicted strain, and no further substantial strain change was expected for each state. We consider that the separation of the impact of strain on dv/v, induced by drying shrinkage, and the pure impact of water saturation of the concrete's pore system, induced by varying relative humidity, is unnecessary because the two effects always occur simultaneously.

Interpreting the results in Figure 12, we find a steady decrease of dv/v for an increase of water in the pore system up to 80 % RH. This trend is disrupted for the highest saturation at 95 % RH, where dv/v increases compared to the previous state. Additionally, the evaluation of the quality factor Q yielded a steady increase in attenuation for all investigated states as relative humidity increases (Figure 15). In addition, the ultrasonic signal correlation between 65 % RH and 95 % RH shows a relatively small CC of approximately 0.6. In contrast, all other signals below 80 % RH as used in plotting Figure 12, have a $CC > 0.8$. Investigating the ultrasonic waveforms, we found a strong attenuation of the shear wave for high saturation, which we explain as due to the presence of fully saturated pores. As coda waves are dominated by S waves [138], the strong attenuation may induce significant waveform distortion such that the stretching technique suffers from cycle skipping. Figure 10 shows a jump in the CC values between 95 % RH and 80 % RH that is not observed for the later step-decrease in relative humidity. CWI often measures confidence using CC values. As a result of this high uncertainty in interpreting the dv/v measurements at such high saturation, we excluded the 95 % RH state from the polynomial model in Figure 13. Typically in seismological research, the contributions of the temperature and relative humidity effects are summed in a linear combination to explain dv/v [111, 139], though these relations have strong spatial heterogeneity [114, 113]. Here we noticed a change in sensitivity to temperature depending on the relative humidity. Hence, we recast the problem using a polynomial regression model to perform linear regression on the polynomial features. The polynomial features are the model variables, i.e., the water-to-cement ratio, the temperature, relative humidity, and their powers up to a polynomial degree of 3. Due to the specificity of the experimental setup, we are not interpreting the absolute values of the coefficients. We propose a way to solve for relative humidity and temperature given the two seismic observables dv/v and Q. Furthermore, we can interpret that the relations are not strictly linear if one compares materials with different porous properties. As the model is an analytical model, an analysis of the sensitivity of the relative velocity change to the ambient conditions and the water-to-cement ratio can be performed. The model can readily compensate for ambient conditions when performing damage detection in concrete structures under non-standard weather conditions. It should be noted that the model cannot be extrapolated and must only be used within the limits of the experimental measurements, especially since we noted a strong sensitivity of dv/v to large RH values. As an alternative to the approach used here, we tested a Gaussian Process Regression and found a good fit. However, the lack of interpretability hindered the characterization of the physical processes.

3.5 Conclusions

The objective of our study was to derive relationships between the ambient variables temperature and relative humidity, and changes in ultrasonic signals in concrete evaluated using Coda wave interferometry and attenuation measurements. Thanks to our experimental design, we were first able to establish one-dimensional relationships for the sole variation of either temperature or relative humidity.

We confirmed the findings of previous studies [107, 137], which demonstrated an approximately linear effect of temperature variation on the velocity variation of ultrasonic signals. Furthermore, we extended these investigations to include the impact of relative humidity variations. We were able to show that the stretching method is an appropriate and sensitive method for measuring the velocity variation of the signals for state changes between 30 % RH and 80 % RH. However, our measurements only had a low confidence level based on the signal correlation coefficient for the state at 95 % RH, where the concrete's pore system is highly saturated with water. We hypothesized that the signal perturbations were too large and caused decreasing confidence in CWI for the highly saturated state. We substantiated this hypothesis by introducing the quality factor to study the signal attenuation, particularly sensitive to microstructure and its damage. Its evaluation showed a decreasing value of the quality factor (e.g., an increase in attenuation) for a higher degree of relative humidity. The steady trend for the analysis of Q finally reinforced the presumed low confidence for the particular data point at very high saturation.

As we observe changes in both temperature and relative humidity in natural settings, we extended our investigations with a polynomial model that accounts for the simultaneous variation of both variables. Furthermore, we embedded the water-to-cement ratio into our model, which defines the characteristics of the concrete pore system. For a higher w/c ratio, we expect a greater pore volume and, therefore, faster dynamics regarding the fluid transport processes in the material. We could confirm this assumption by explicitly analyzing the velocity variation of the ultrasonic signal, which showed greater sensitivity to temperature and relative humidity changes for a greater water-to-cement ratio (i.e., greater pore space). Considering all three variables leads to our model being of the third order.

Evaluating the changes in phase and amplitude of ultrasonic signals in concrete allows the detection of even weak changes induced by ambient conditions. To subsequently detect irreversible changes or damage in the framework of an early-warning system, we conclude that the proportion of velocity and attenuation variation changes, generated by ambient variables, can be characterized and hence compensated. For the separation of temperature-induced changes, we consider the simultaneous monitoring of T using embedded sensors to be the most elegant way of controlling its impact on dv/v, while the observation over a period that covers the typical temperature range, may be sufficient for deriving the $dv/v(T)$ relationship because of its linearity. Regarding the impact of the relative humidity on dv/v, the change is slow and only visible over a long-term perspective. However, changes in Q are significant, and attenuation may be used instead of dv/v to compensate for relative humidity. Therefore, we propose an ambient seismic field approach to monitoring the internal state of concrete that will provide a more accurate assessment if or when structural damage occurs.

Appendix Chapter 3 – Polynomial Model Coefficients

Table 5: List of coefficients of the third order polynomial obtained by polynomial regression.

n	$a[ijk]$	i	j	k
0	-1.235967e-01	0	0	0
1	9.621853e-02	1	0	0
2	3.631449e-04	0	1	0
3	5.107886e-03	0	0	1
4	1.344006e-01	2	0	0
5	1.380768e-04	1	1	0
6	-2.651369e-03	1	0	1
7	-1.546802e-06	0	2	0
8	-2.261435e-06	0	1	1
9	-7.619478e-05	0	0	2
10	1.151416e-01	3	0	0
11	4.693153e-05	2	1	0
12	-4.163067e-03	2	0	1
13	-7.605707e-06	1	2	0
14	5.858475e-06	1	1	1
15	2.860905e-05	1	0	2
16	1.809416e-08	0	3	0
17	9.979734e-08	0	2	1
18	-3.073448e-08	0	1	2
19	3.765474e-07	0	0	3

4 Impact of External Mechanical Loads on Coda Waves in Concrete [2]

Abstract - *During their life span, concrete structures interact with many kinds of external mechanical loads. Most of these loads are considered in advance and result in reversible deformations. Nevertheless, some of the loads cause irreversible, sometimes unnoticed changes below the macroscopic scale depending on the type and dimension of the impact. As the functionality of concrete structures is often relevant to safety and society, their condition must be known and, therefore, assessed on a regular basis. Out of the spectrum of non-destructive monitoring methods, Coda Wave Interferometry using embedded ultrasonic sensors is one particularly sensitive technique to evaluate changes to heterogeneous media. However, there are various influences on Coda waves in concrete, and the interpretation of their superimposed effect is ambiguous. In this study, we quantify the relations of uniaxial compression and uniaxial tension on Coda waves propagating in normal concrete. We found that both the signal correlation of ultrasonic signals as well as their velocity variation directly reflect the stress change in concrete structures in a laboratory environment. For the linear elastic range up to 30 % of the strength, we calculated a velocity variation of $-0.97‰$ / MPa for compression and 0.33 % / MPa for tension using linear regression. In addition, these parameters revealed even weak irreversible changes after removal of the load. Furthermore, we show the time-dependent effects of shrinkage and creep on Coda waves by providing the development of the signal parameters over time during half a year together with creep recovery. Our observations showed that time-dependent material changes must be taken into account for any comparison of ultrasonic signals that are far apart in time. The study's results demonstrate how Coda Wave Interferometry is capable of monitoring stress changes and detecting even small-size microstructural changes. By indicating the stated relations and their separation from further impacts, e.g., temperature and moisture, we anticipate our study to contribute to the qualification of Coda Wave Interferometry for its application as an early-warning system for concrete structures.*

[2] This article appeared in its original form in the peer-reviewed, open access journal *materials* **15** (16), 5482 on August 9, 2022 [103]. It is adpoted verbatim in this chapter. The authors' work share is listed in Appendix 2.

4.1 Introduction

Reinforced concrete is used for a large part of our built environment today [140], especially in engineering structures such as our infrastructure. Besides the essential safety aspect, reliability and permanent availability are major objectives regarding these structures' design and maintenance. Higher demands, especially due to rising traffic volume and heavy load traffic, and the simultaneous aging of the structures, necessitate increased attention to maintenance and monitoring to avoid disruptive effects such as local material failure or even a devastating collapse. Regular inspections, in most cases based on visual examination or tap tests, ensure structural integrity and operability. On the downside, they are only able to represent the structure's state at accessible areas at one point in time, whereas monitoring methods, usually from the field of non-destructive testing (NDT), complement the spectrum of available tools for safety assurance. Reinforced concrete, however, poses a special challenge for NDT techniques due to the material's heterogeneity, which complicates the separation between naturally present scatterers and undesired material changes, or damage. To this end, the evaluation of ultrasonic signals by means of Coda Wave Interferometry (CWI) is a promising and exceptionally sensitive method to detect even weak changes in the material and has been subject to previous studies on both the laboratory [89, 101, 104, 141, 142], and the structural scale [109, 32, 34, 143]. It is suitable for application as a permanent monitoring system with the focus on the early warning of microcrack initiation, and possibly providing the impulse for further in-depth inspections at alarming locations. With this study, we quantify the effects of uniaxial compression, uniaxial tension, shrinkage and creep of a normal concrete on signal changes in the Coda that we describe by means of correlation coefficients and velocity variations of collected ultrasonic signals.

4.2 State of the Art

4.2.1 Coda Wave Interferometry

Ultrasonic (US) waves in inhomogeneous media are scattered, depending on the wave properties such as frequency, and the size, shape, and type of the inhomogeneities. The effect of such scattering on ultrasonic measurements is shown in Figure 16. Although we record a distinct first arrival, which is used in classic US non-destructive testing (NDT) methods, a significant portion of the energy emitted at the source arrives at the receiver later than this first arrival. This long tail of the measurement is called Coda. Although the origin of Coda wave-based methods is found in the field of seismology, the analysis of such waves by Coda Wave Interferometry has been applied to monitoring of solids in recent decades (see e.g., [144, 17, 145, 31]). Using CWI methods, evaluation of consecutive ultrasound measurements enables detecting weak changes to the signal. This technique is based on the fact that the medium is acting as an interferometer, merging the scattered waves in the receiver position [17]. As long as the medium (interferometer) does not change, repeated measurements will record identical signals. Changes to the medium as a result of local or global impacts directly affect the recorded signals. If the change is between source and receiver, it will be visible in the first arrival. Changes outside the direct path are only affecting the scattered waves and can, therefore, only be detected using CWI analysis of volumetric ultrasound information (Figure 16). Generally, CWI methods take advantage of the accumulation of weak changes over a great number of wave paths between a sender-receiver pair that are caused by the heterogeneities. Multiple methods for CWI analysis have been proposed in recent decades [17, 86, 31], all based on the calculation of the signal correlation coefficient (CC). This coefficient determines the similarity of a reference signal $u(t)$ in comparison to a signal recorded after a perturbation in the material $\tilde{u}(t)$. In this work, we will use the stretching technique (see [86]) to determine apparent relative velocity changes (dv/v) in the medium. This method applies different stretching factors $\varepsilon = dv/v$ to the reference signal and determines the ε maximizing CC:

$$CC(\varepsilon) = \frac{\int_{t_1}^{t_2} u(t)\tilde{u}(t(1-\varepsilon))\,dt}{\sqrt{\int_{t_1}^{t_2} u^2(t)\,dt \int_{t_1}^{t_2} \tilde{u}^2(t(1-\varepsilon))\,dt}} \tag{4.1}$$

With this method, two parameters, $\varepsilon = dv/v$ and CC can be extracted from the ultrasonic coda in the time interval $[t_1, t_2]$ and be used to analyze and track global and local material changes. The reference measurement $u(t)$ can either be fixed for an entire experiment or change during the course of it if the differences between $u(t)$ and $\tilde{u(t)}$ become too large (e.g., $CC \leq 0.7$) due to damage or material property changes [119]. Both the fixed reference and the changing reference method can be applied to relate the velocity change in a monitoring experiment to the initial state of the specimen. Although for the fixed reference method this is straightforward, a stepwise change in reference requires cumulative calculation of the total velocity change at every step. When switching from fixed reference to the stepwise method, CC can no longer be used for interpretation.

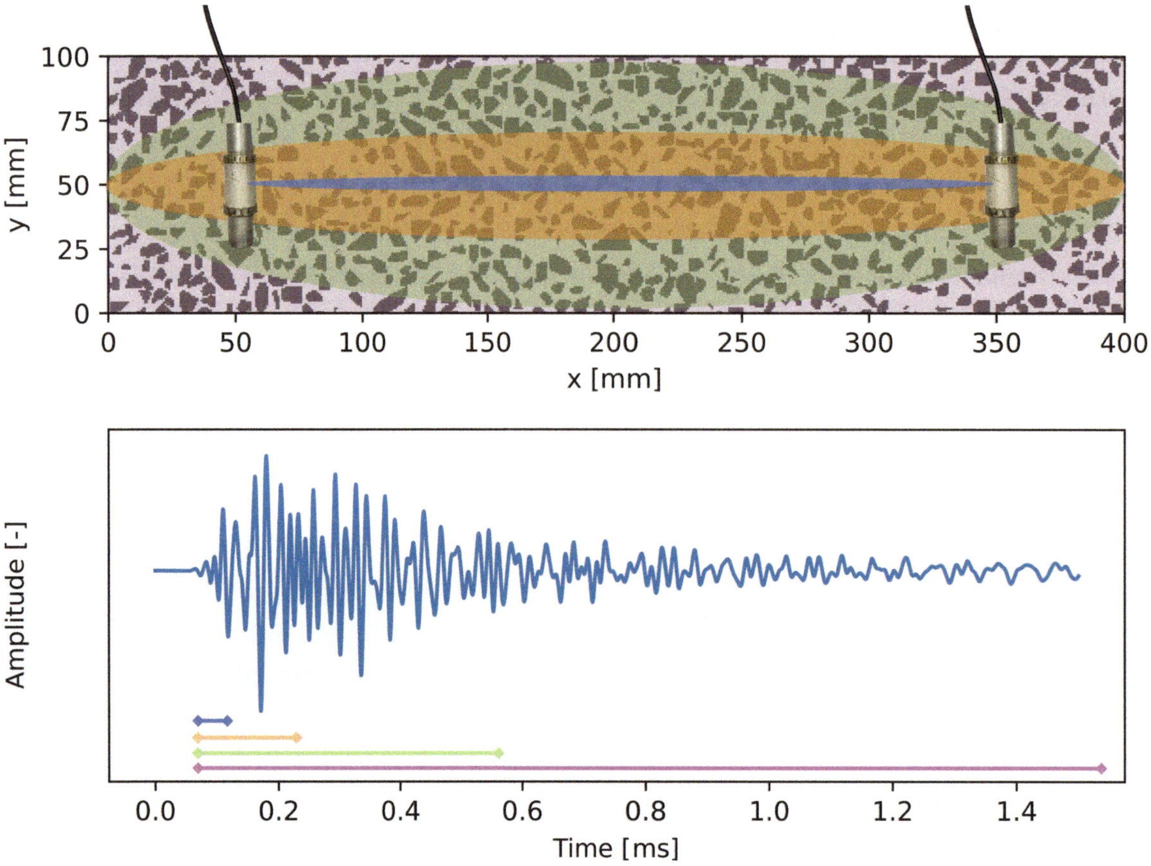

*Figure 16: Ultrasonic recording (**bottom**) and the schematic explanation of CWI with two US transducers in a concrete model (**top**): when analyzing only the first arriving waves, the measurements are sensitive only to the area around the direct path between source and receiver (blue). As more scattered waves are included, the area increases (orange, green). Referring to the entire signal (purple), scattered waves from the entire medium are recorded at the receiver.*

4.2.2 Mechanical Loading of Concrete

Concrete is a complex heterogeneous material consisting of coarse aggregates and a system of air pores embedded into a cementitious mortar matrix. Depending on the design case and objective, concrete can be investigated on different scales, whereas most structural and material design models consider the material on the macro level as a homogeneous material. However, evaluation of crack formation starts with pre-existing defects on the micro level, followed by a multi-layered process of damage evolution [146]. From the monitoring perspective, it is most ideal to detect weak changes in the material already on the microstructural level and use the collected information as a precursor for material failure on the above-lying scales. Besides the impact of environmental conditions such as temperature or moisture, changes are in many cases caused by mechanical loads, i.e., compression, tension, or multidimensional stress states. We show the crack formation process for a normal concrete for the uniaxial compression and tension cases in Figure 17, which we cover in our experiments. Preexistent micro-cracks evolve primarily in tensile sections along the comparably weaker boundary between the cement matrix and single aggregates, the Interfacial Transition Zone [147]. For the compression case with much higher strength, stress trajectories appear between adjacent aggregates [148], whereas for tension, a single separating crack develops [149]. Besides direct impacts such as short-term loading, time-dependent phenomena such as shrinkage and creep are characteristic for concrete. Shrinkage causes a reduction in the material's volume and is affected by the ambient humidity, the component geometry, the drying process, and the concrete mix. It is a superimposed process of drying shrinkage due to moisture loss to the environment and autogenous shrinkage due to the internal reaction of water during the hydration from fresh to hardened concrete [150]. Although deformations as a consequence of shrinkage are generally load-independent, creep deformations are considered to depend primarily on the load level and its duration of exposition but also on moisture migration and micro crack formation [151].

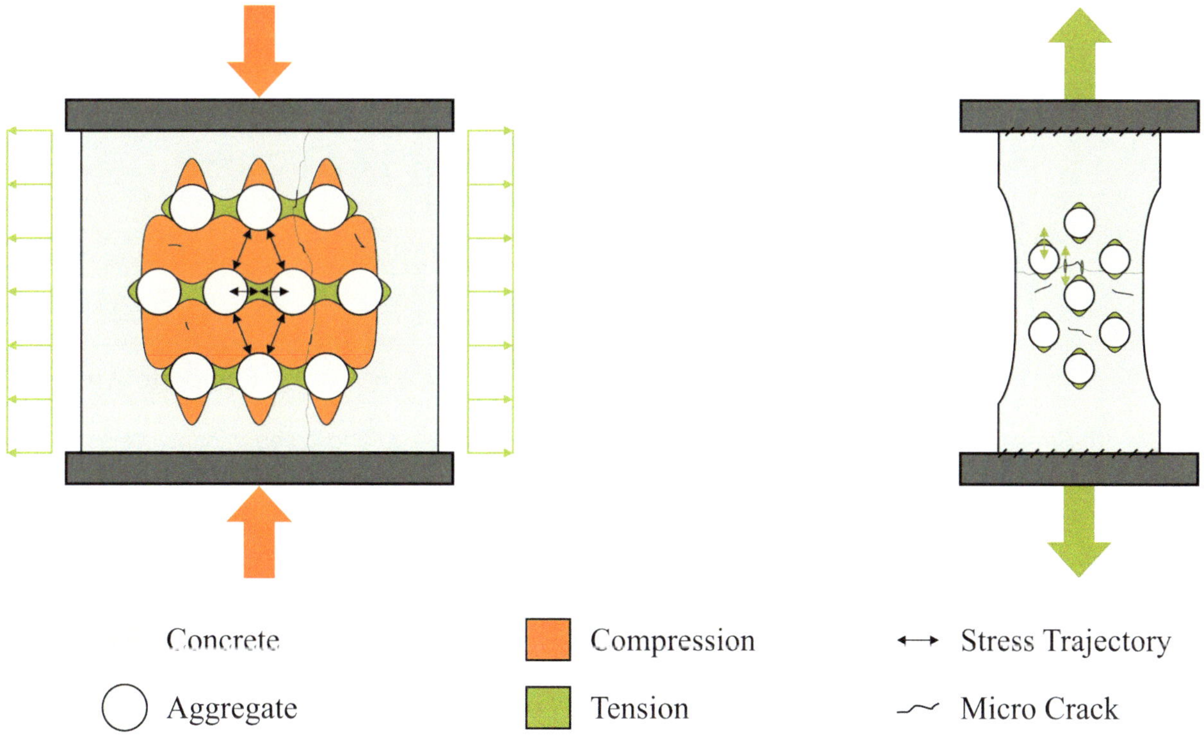

*Figure 17: Uniaxial loading of normal concrete specimens, where the cement paste's Young's modulus is lower in comparison to the aggregate's Young's modulus. The experimental setups for both compression (**left**), and tension (**right**), correspond to the experiments in Section 4.3. The compression case is shown with the assumption of no friction at the contact surface between press and specimen.*

The immediate strain $\varepsilon_{ci}(t_0)$, together with a plastic term $\varepsilon_{pl}(t_0)$, is complemented by two terms that account for shrinkage $\varepsilon_{cs}(t_s)$ and creep $\varepsilon_{cc}(t)$ for a constant compressive stress $\sigma(t_0)$ during the loading period between t_0 and t_e. Equation (4.2) [24] summarizes the total strain $\varepsilon_c(t)$ (Figure 18) for a load application at t_0 at constant environmental conditions.

$$\varepsilon_c(t) = \varepsilon_{cs}(t) + \varepsilon_{ci}(t_0) + \varepsilon_{pl}(t_0) + \varepsilon_{cc}(t, t_0) \tag{4.2}$$

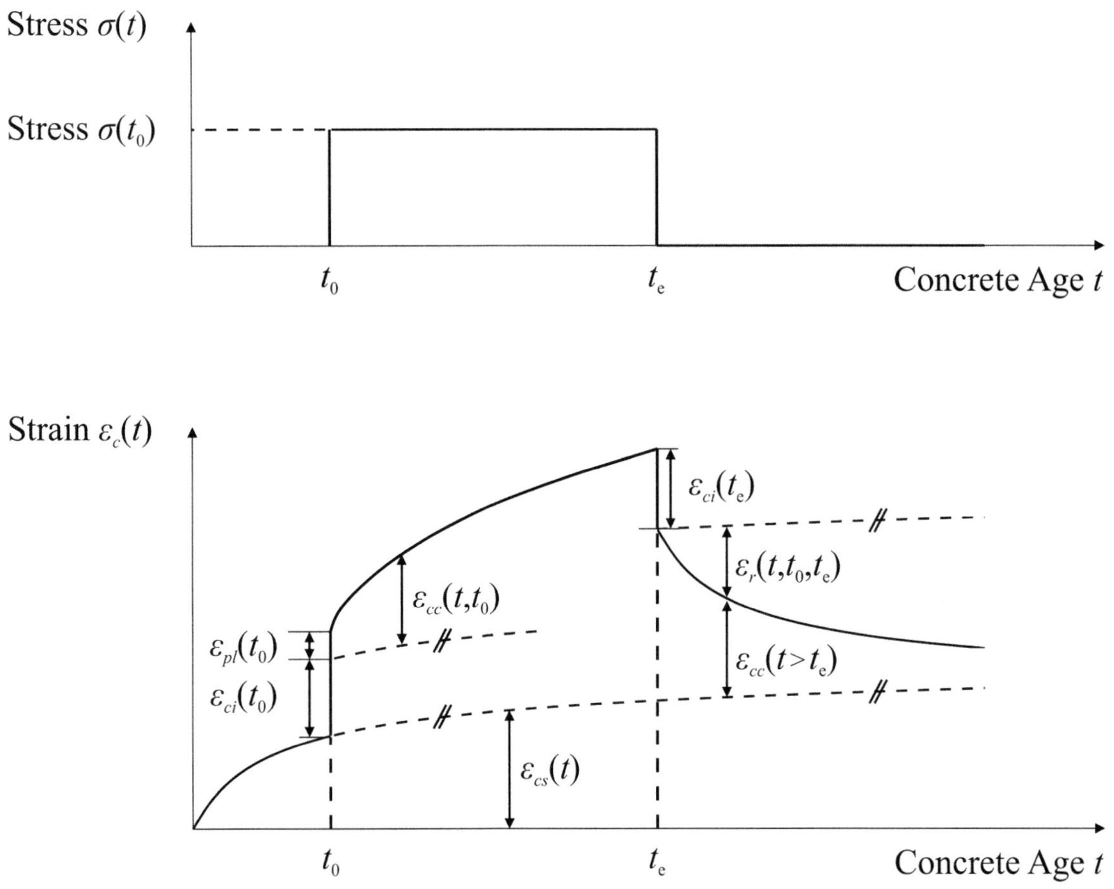

*Figure 18: Stress (**top**) and corresponding strain (**bottom**) for shrinkage and creep in concrete as a function of concrete age t. The characteristic points t_0 and t_e specify loading and unloading with a constant load $\sigma(t_0)$.*

4.3 Materials and Methods

4.3.1 Experimental Setup

We used an identical concrete mix for all experiments. Table 6 shows the concrete raw materials and their composition. Pre-tests for the Young's modulus yielded 84.6 GPa for the mainly quartzitic crushed aggregates and 27.1 GPa for the hardened mortar. The Poisson's Ratio was 0.12 for the aggregates and 0.19 for the mortar. The compressive strength was 368.0 MPa for the aggregates and 80.3 MPa for the mortar. A standardized pressure method test yielded an air void content of 3.9 % in the fresh concrete. The polycarboxylatether-based superplasticizer was added to the mixing water to improve the concrete's workability and reduce the cement paste viscosity to enhance the sensor-to-concrete coupling condition. The characteristic concrete parameters were additionally modeled [24], particularly considering the concrete model's damage evolution [23]. After production, the specimens were cured at 20 °C under water for 7 days and restored to 65 % RH and 20 °C. We started each experiment in this study on the 28th day after concrete production.

Table 6: Concrete raw materials and their composition for all experiments in this study, where the aggregates are primarily quartzitic and crushed.

Cement Content	**Water / Cement**	**Aggregate Size**			**Superplasticizer**
CEM I 42.5 R	w/c	0/2 mm	2/5 mm	5/8 mm	BASF MasterEase 3880
350 kg/m³	0.45	46.5 %	34.0 %	19.5 % *	2.5 % **

* Contains 3.0 % oversized grains 8/12 mm.

* As a mass fraction in % of the cement content.

4.3.2 Compression Experiment

For the compression test, we used three cuboid specimens with an edge length of 400 mm and side lengths of 100 mm with a transducer distance of 300 mm symmetrical to the center axis (Figure 19). We decoupled the specimen on the sawn end faces by a polytetrafluorethylen (PTFE) film from the hydraulic press to minimize friction, thus avoiding a multiaxial stress state next to the faces. In advance, we evaluated the average compressive strength f_c of three geometrically identical samples without embedded transducers, which was 42.9 MPa. Subsequently, we applied a compressive load to the actual specimens with a loading rate of 50 N/s while recording ultrasonic signals between the embedded transducers every 10 s. In advance, all specimens were loaded at $5\%f_c$ to dissolve initial internal stress. We defined different maximum loads for each specimen, i.e., 100 %, 60 %, and 30 % of the compressive strength f_c. Having reached the maximum load, we unloaded the sound specimens again.

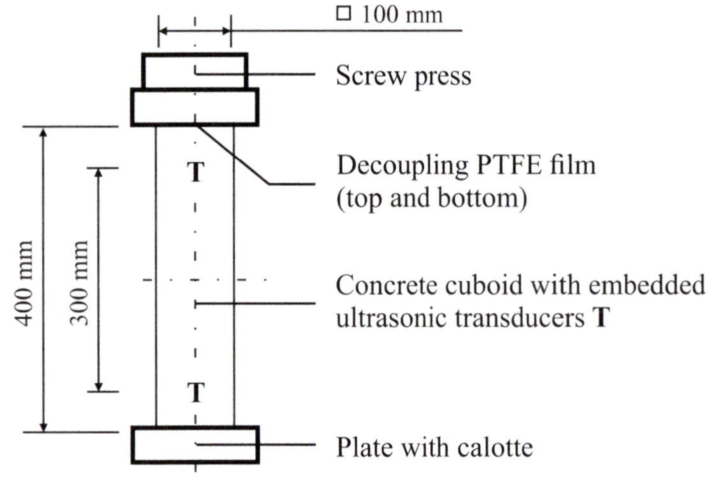

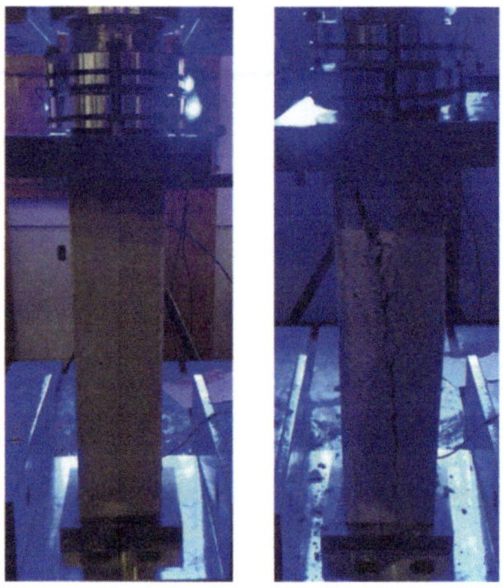

*Figure 19: Setup for the uniaxial compression experiment: (**left**) sketch of the cuboid concrete specimens with their dimensions, (**center**) picture of the experiment, and (**right**) picture after failure. A pre-test with three identical specimens yielded a compressive strength of 42.9 MPa.*

4.3.3 Tension Experiment

For the centric tension test, we used three cylindrical bone-shaped specimens with a length of 300 mm and a decreasing cross section [149] to define a predetermined fracture section (Figure 20). The transducer distance was 198 mm symmetrical to the center axis. The specimens were fastened at the tensile machine using the two-component adhesive *Hilti HIT-RE 500 V3* having cut the end faces in a chessboard pattern to improve the strength of the joint. The tensile strength was determined to be 3.0 MPa beforehand from three identical samples. The subsequent loading rate was 5 N/s while recording every 10 s up to maximum loads of 100 %, 60 %, and 30 % of the tensile strength f_{ctm}. Like the compression test, we unloaded the sound specimens again.

4.3.4 Shrinkage and Creep Experiment

The specimens for the shrinkage and creep tests were cuboids, geometrically equal to the ones from the compression test. The edge length was 400 mm and side lengths were 100 mm with a transducer distance of 300 mm symmetrical to the specimen's center axis (Figure 21). The sawn end faces were decoupled by a PTFE film from the steel frames in which a hydraulic system instantaneously compressed the specimens at different constant loads of 11 %, 18 %, and 33 % of the previously determined compressive strength. In addition, we investigated a sealed specimen at a load of 33 % and an unloaded specimen to monitor pure shrinkage. We applied the load for the duration of half a year and continued monitoring for another 50 days after unloading to observe creep recovery.

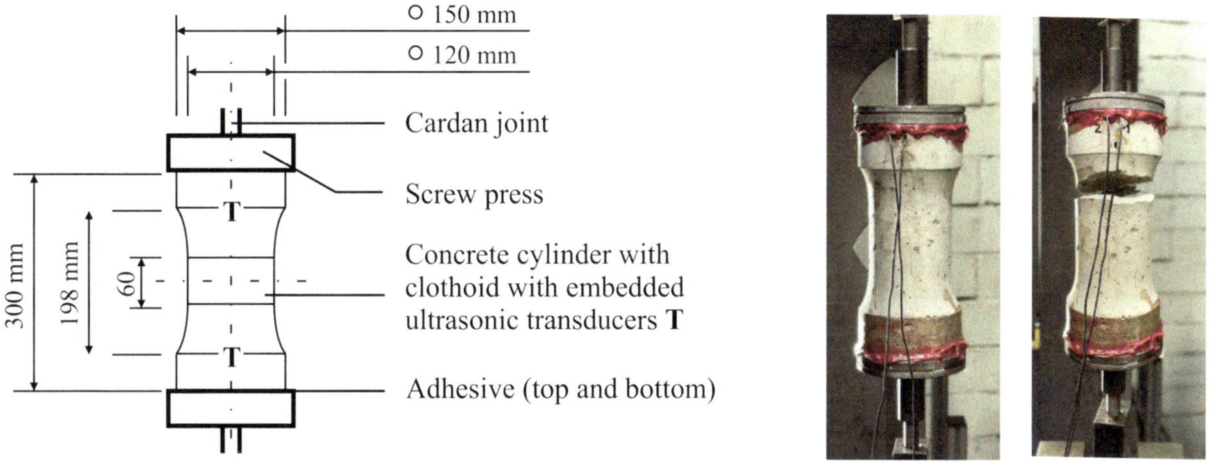

*Figure 20: Setup for the uniaxial tension experiment: (**left**) sketch of the cylindrical concrete specimens with their dimensions, (**center**) picture during the experiment, and (**right**) picture after failure. A pre-test with three identical specimens yielded a tensile strength of 3.0 MPa.*

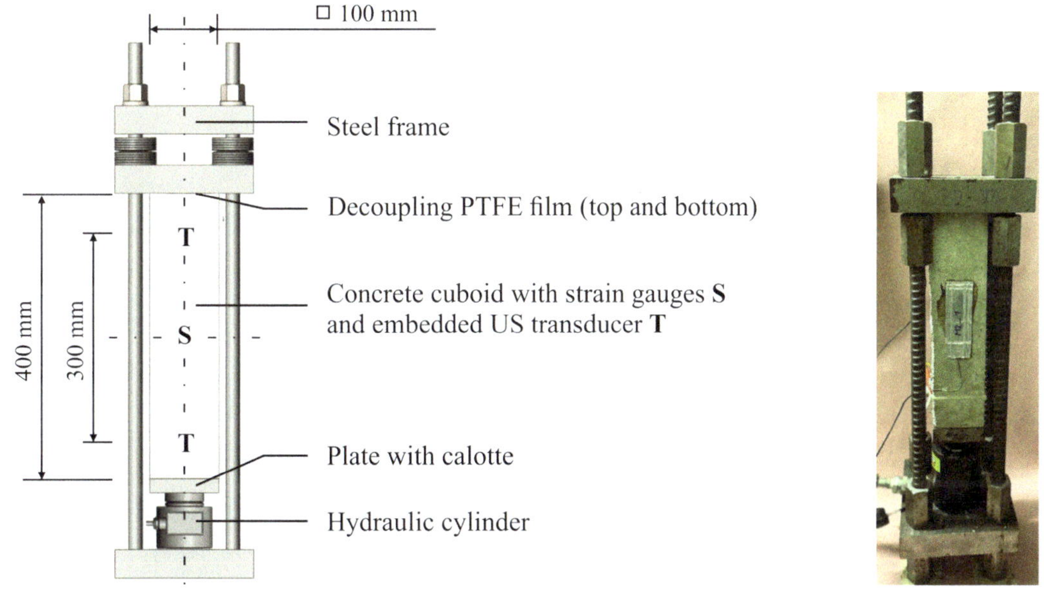

*Figure 21: Setup for the shrinkage and creep experiment: (**left**) sketch of the cuboid concrete specimens with their dimensions, loaded within a steel frame, and (**right**) picture of the experiment.*

The normal strain of every specimen was individually tracked by a system of strain gauges in a temperature-compensated circuit of two active and two passive strain gauges each. All specimens were equipped with embedded piezoelectric transducers (described in [91]) to collect ultrasonic signals. The transducers acted as sender-receiver pairs with a center frequency of 60 kHz. The measurement system (described in [127]) recorded the resulting signal with a sampling rate of 1 MHz and 12,000 samples with a resolution of 14 bits. We subsequently compared the signals without further averaging.

4.4 Results

For each experiment, we evaluated the collected ultrasonic signals using Coda Wave Interferometry methods. By comparing every signal to a fixed, non-perturbed reference signal at an unloaded state, we present the velocity variations dv/v together with the signal correlation coefficient CC according to Equation (4.1) during the individual loading scenarios. Positive values of dv/v indicate a velocity decrease and vice versa. All ultrasound measurements were preprocessed before application of the CWI algorithm. Preprocessing included offset compensation, pretrigger and crosstalk removal and frequency filtering. We refer to [119] for the detailed processing procedure. CWI parameter changes as a consequence of either compressive or tensile stress may be interpreted as calibration curves, whereas the development of CWI parameters over time due to the phenomena shrinkage and creep are referred to as concrete-specific features.

4.4.1 Uniaxial Compression

The compressive loading scenario is presented in Figure 22 together with the velocity variation dv/v and the signal correlation coefficient CC for a fixed reference signal in an unloaded state. The loading and unloading rate was 50 N/s and equal for all specimens. One specimen was loaded up to failure ($\sigma_{max} = f_c$), which caused maximum velocity variation at failure of approximately 11 %. The other specimens were loaded up to 30 and 60 % of the maximum load f_c. The maximum load was held for 5 min, followed by unloading. During the whole experiment, we left the press connected to the specimen by applying a minimum load of 1 $\% f_c$ to keep the contact condition of the press unchanged. Ultrasonic signals were collected at a rate of 0.1 Hz. As the experiment covers not only weak changes in the stress state, using one fixed reference for the entire experiment may cause discontinuities of velocity perturbations due to great signal variations in comparison to an unloaded state. We considered this method-specific phenomenon by establishing an additional evaluation technique picking a new reference signal each time the overall signal correlation coefficient drops below a threshold, which was defined as 0.7 for this analysis. Therefore, we ensure keeping the signal changes within the thus generated windows small enough to satisfy the principles of Coda Wave Interferometry regarding the sensitive evaluation of weak changes to the material. Up to a compressive stress of 6.4 MPa, equaling $0.15 f_c$, we observe insignificant deviations between the two methods for the development of dv/v. For a growing stress, deviations increasingly change and the gradient switches sign for the two specimens $\sigma_{max} = 0.30 f_c$

and $\sigma_{max} = 0.60 f_c$. Numerical instabilities are observed particularly for the specimen $\sigma_{max} = f_c$. Analysis of the velocity variation yielded similar development over time, or equally load. We observed a steady signal compression for increasing loads up to approximately 25.6 MPa, equaling 60 %f_c, resulting in a maximum dv/v of –1.6 %. For further loading, the velocity variation gradient switched sign and increasingly rose, indicating evolving damage in the material. For the two non-failed specimens, dv/v turned positive while unloading and reached a maximum of 0.6 % for $\sigma_{max} = 0.30 f_c$ and 0.9 % for $\sigma_{max} = 0.60 f_c$, indicating irreversible material changes after the experiment. A linear regression analysis of the three specimens up to $0.30 f_c$ yielded a velocity variation increase of -0.97 ‰/MPa which is comparable to the slope in [46]. The coefficient of determination for this analysis is 76.6 %.

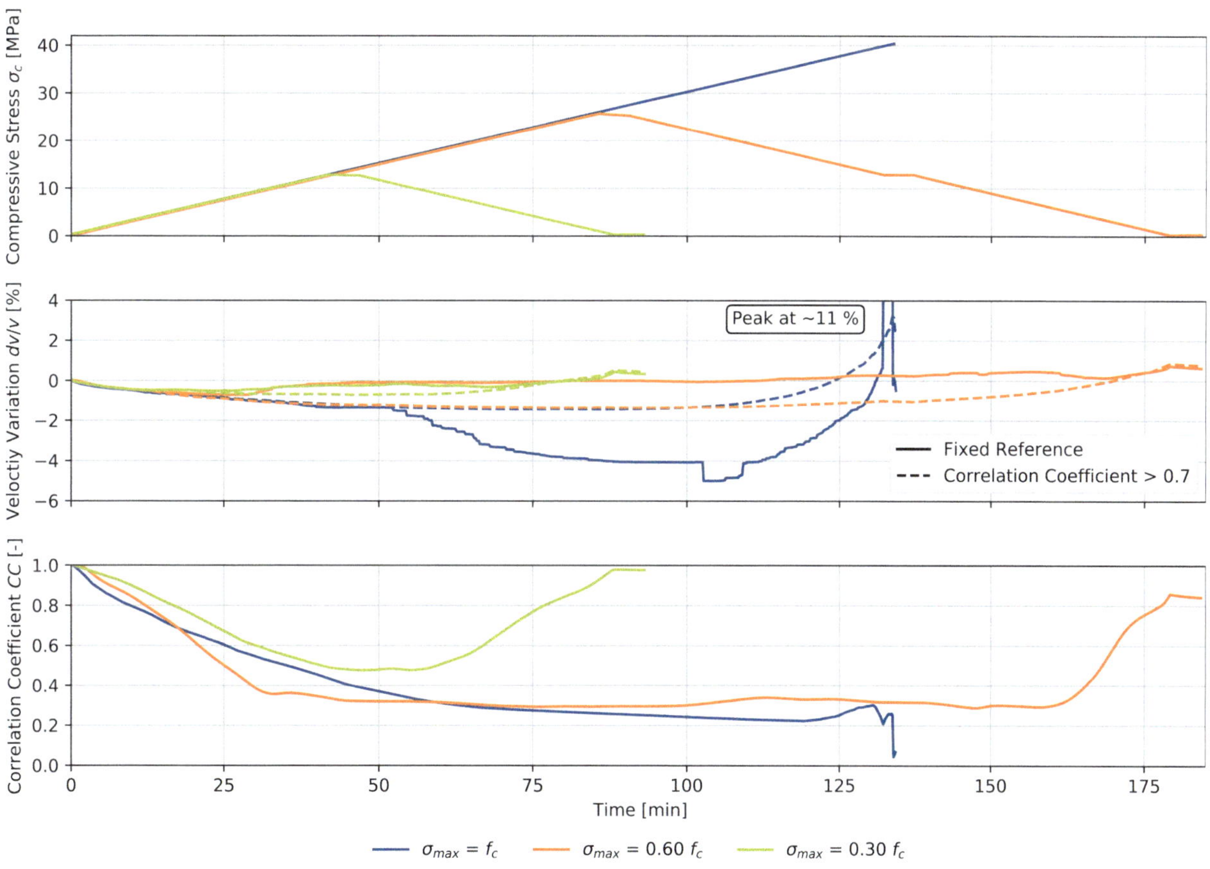

Figure 22: Uniaxial compression experiment for three specimens, referring to the setup in Figure 19 with different maximum stresses: compressive stress σ_c, velocity variation dv/v and signal correlation coefficient CC over time.

Analysis of the signal correlation coefficient development also yielded similar results for all specimens up to 12.8 MPa, equaling 30 %f_c, where CC expectedly drops for increasing loads. For CC below 0.4 to 0.3, its gradient is attenuated, and does not drop below 0.2 before material failure. Regarding the two unloaded specimens, we observed an increase in CC, but not a complete restoration of the original signal at the unloaded state. The remaining offset was approximately 0.03 for $\sigma_{max} = 0.30 f_c$ and 0.15 for $\sigma_{max} = 0.60 f_c$. As environmental and bearing conditions did not change during the experiment, the remaining offset is connected to the internal material damage of the specimen, which is significantly larger for higher loads, and corresponds to the observations referring to the signals' velocity variation dv/v. Both parameters were translated into the compressive stress domain during the loading process. We show the results in Figure 23. The velocity variation is plotted in accordance with both methods for one fixed reference (top) and switching reference signals for windows with CC > 0.7 (bottom). As both ultrasonic signals and the stress have been measured in the time domain, CWI parameters are linearly interpolated for the respective stress. For a measuring rate of the load of 10 Hz and a loading rate of 50 N/s, we specify a resolution of < 0.25 MPa. Evaluating CC only above the threshold 0.7 yields a steady development of dv/v for increasing loads up to $\sigma_{max} = 0.60 f_c$, whereas dv/v is not completely steady, in particular for sections with major changes to the ultrasonic signals. Considering building codes [135], the elastic modulus of concrete can be estimated referring to its strength, depending on the used aggregates. For the concrete with quartzitic aggregates, the secant modulus between $\sigma_c = 0$ and $40\,\% f_{cm}$ is applied for determining the elastic modulus, where f_{cm} is the medium compressive strength. Referring to the standards, the relation between compressive stress and the CWI parameters is nearly reversible within this range, which has been confirmed up to 30 %f_c in this experiment through an approximate restoration of the original values of dv/v and CC after unloading. However, deviations as a consequence of damage evolution on a microstructural level must be expected due to the heterogeneous nature of concrete.

4.4.2 Uniaxial Tension

The centric tensile loading scenario is presented in Figure 24 together with the velocity variation dv/v and the signal correlation coefficient CC. The indicated tensile stress refers to the center circular cross section with a diameter of 120 mm. The loading rate was 5 N/s and equal for all specimens. As in the compression experiment, one specimen was loaded up to failure, whereas two specimens were loaded to 30 % and 60 % of the maximum load f_{ctm}. The maximum load was held for 5 min, followed by unloading. The contact condition during the experiment stayed unchanged due to the adhesive connection. Therefore, no minimum load, as during the compression experiment, was applied. Ultrasonic signals were collected at a rate of 0.1 Hz. In comparison to the compressive

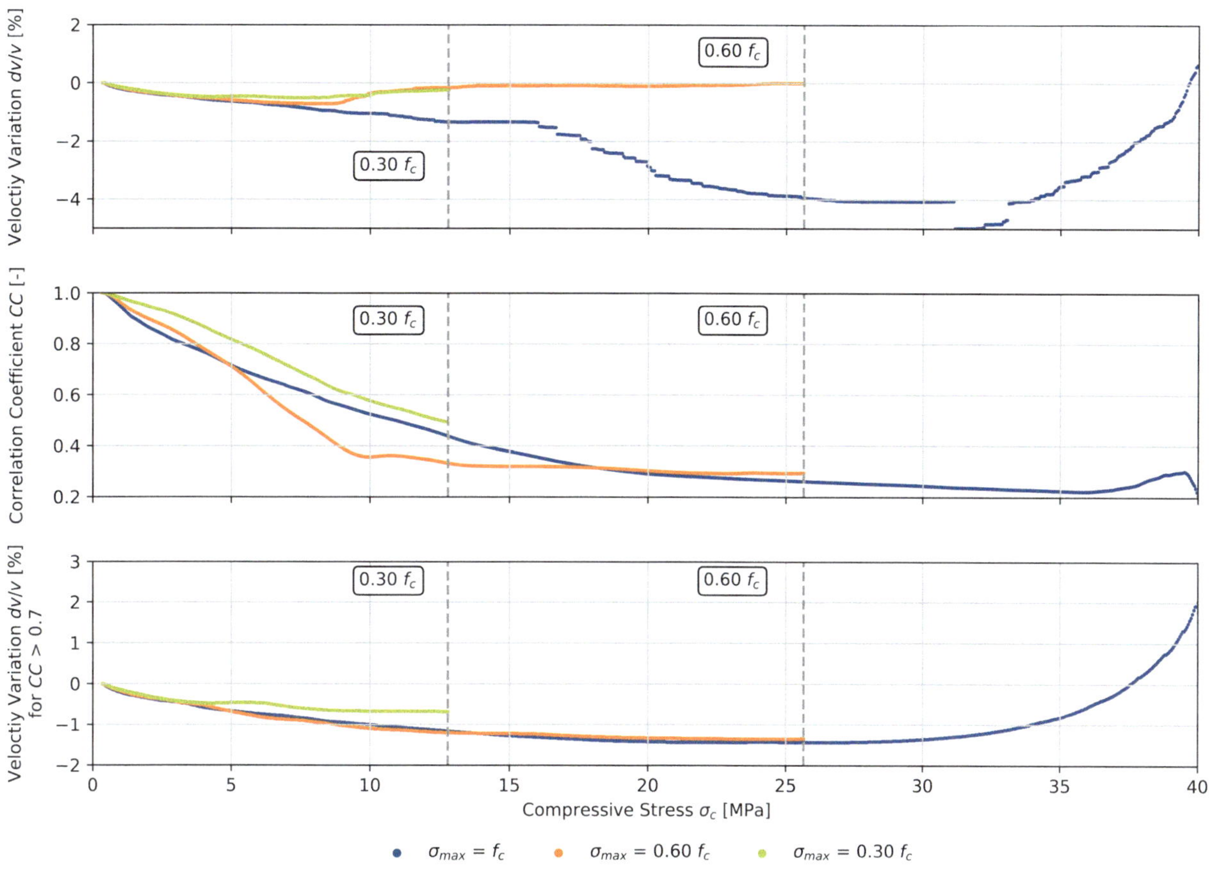

Figure 23: Velocity variation dv/v and signal correlation coefficient CC for three specimens as in Figure 22, translated into the compressive stress domain during the loading process up to 100 %(failure), 60 %, and 30 % of the compressive strength f_c.

loading experiment, the stress range is smaller due to the characteristic behavior of concrete regarding tension. For this concrete mix, the ratio between the tensile strength f_{ctm} and the compressive strength was determined to be around 7.0 %. We show dv/v for a fixed reference signal in an unloaded state only, since CC does not drop as significantly as during compression. Therefore, we consider one fixed reference sufficient to cover the entire experiment.

Analysis of the velocity variation yielded similar forms for all specimens over time, or equally load. Signals were steadily stretched for increasing tensile loads up to 100 %f_{ctm}, or 3.0 MPa. We observed nearly linear development of dv/v, whereas the gradient of one specimen with a maximum load of $\sigma_{max} = 0.60 f_{ctm}$ rapidly decreased above approximately 30 %f_{ctm}. The maximum velocity variation before failure was 0.80 %. Failure was characterized by evolution of one characteristic macro crack. Having reached their respective maxima at 0.36 % for $\sigma_{max} = 0.30 f_{ctm}$ and 0.41 % for $\sigma_{max} = 0.60 f_{ctm}$, dv/v decreased while unloading the specimens. As in the compression experiments, the remaining velocity variation was higher for greater maximum loads, 0.22 % for $\sigma_{max} = 0.30 f_{ctm}$ and 0.37 %

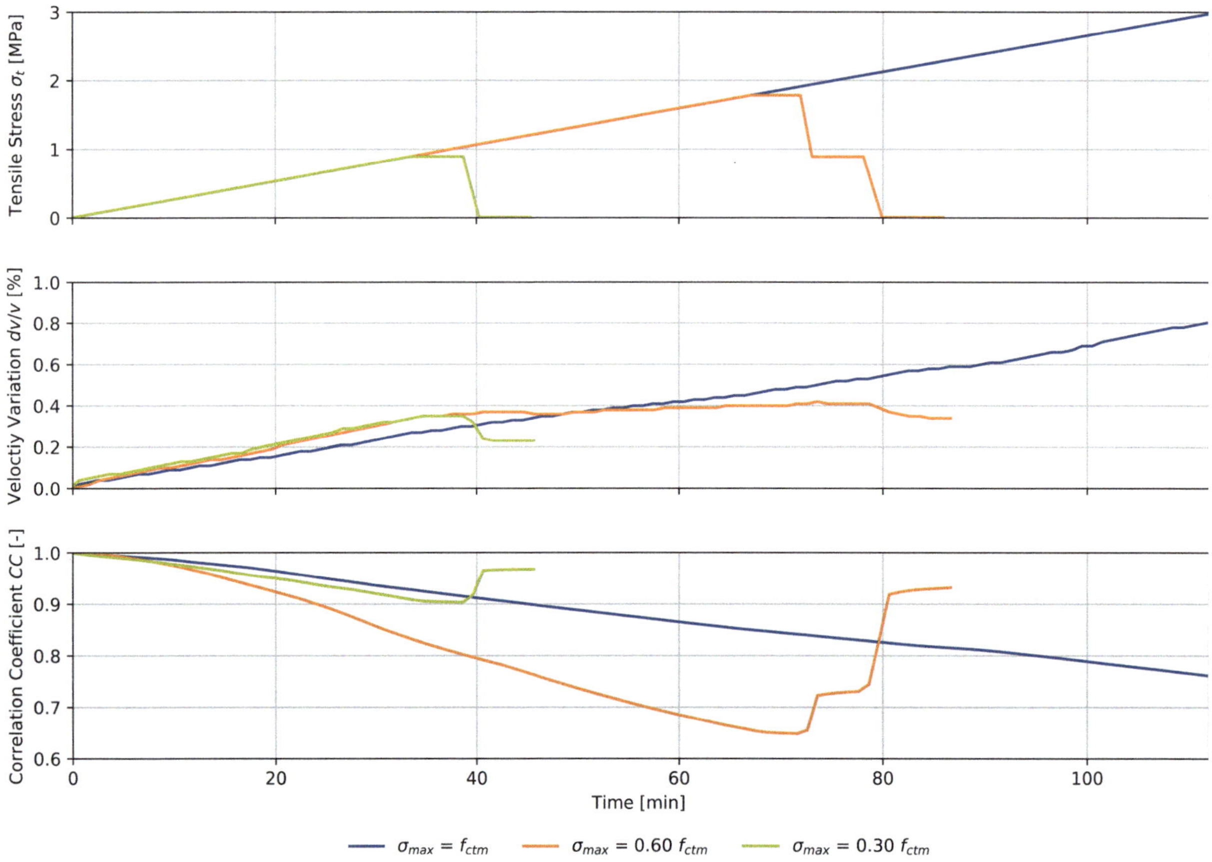

Figure 24: Uniaxial tension experiment for three specimens, referring to the setup in Figure 20 with different maximum stresses: tensile stress σ_t, velocity variation dv/v and signal correlation coefficient CC over time.

$\sigma_{max} = 0.60 f_{ctm}$, again indicating irreversible material changes after the experiment. A linear regression analysis of the three specimens up to $0.30 f_{ctm}$ yielded a velocity variation increase of 0.33 % / MPa. The coefficient of determination for this analysis is 92.2 %. Signal correlation coefficients steadily decreased during the experiment time, or load up to f_{ctm}, whereas deviations between the specimens' CCs also increased for growing tensile stress. The minimum CCs were computed at 0.90 for $\sigma_{max} = 0.30 f_{ctm}$ and 0.66 for $\sigma_{max} = 0.60 f_{ctm}$. After unloading, offsets remained in both cases despite unchanged environmental and bearing conditions. The individual offsets depended on the applied maximum load according to the observations for dv/v. For greater loads, the offset increased, which related to irreversible material changes on a microstructural level. We translated both CWI parameters into the tensile stress domain during the loading process and show the results for the three specimens in Figure 25. Likewise, the compression experiment, CWI parameters are linearly interpolated for the respective stress. For a measuring rate of the load of 10 Hz and a loading rate of 5 N/s, we specify a resolution of < 0.025 MPa. Since the parameters showed a steady increase for greater loads at a constant loading rate, their developments in the tensile stress domain are equivalent to the time domain.

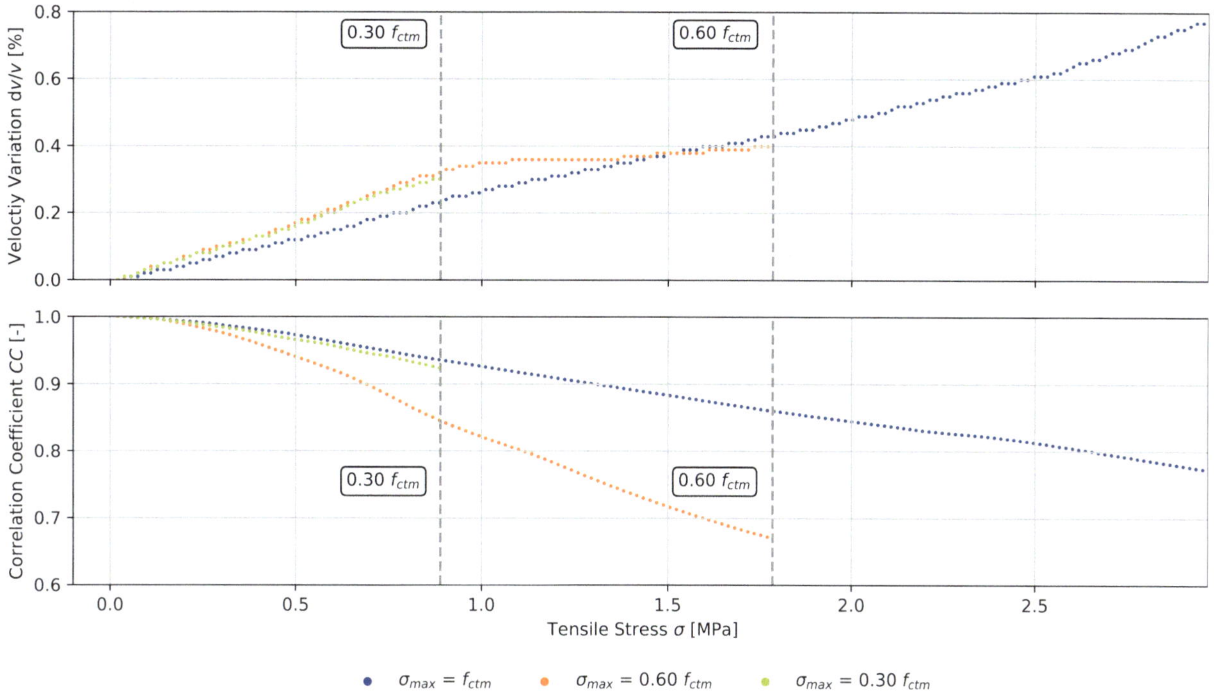

Figure 25: Velocity variation dv/v and signal correlation coefficient CC, both using a fixed reference signal at an unloaded state, for three specimens as in Figure 24, translated into the tensile stress domain during the loading process up to 100 %(failure), 60 %, and 30 % of the compressive strength f_{ctm}.

4.4.3 Shrinkage and Creep

Both processes, shrinkage and creep, are concrete-specific phenomena. In contrast to external load variations, which have a direct impact on the ultrasonic signal, these phenomena cause a characteristic material reaction over time. In this experiment, we show the CWI parameters, together with the measured strain, for pure shrinkage and superimposed creep at different constant loads of $\sigma = [0.11\,\%, 18\,\%, 33\,\%]\ f_c$ in Figure 26. The specimens were loaded for the duration of half a year and unloaded after that period. The spike after 7 days results from a restart of the load-applying hydraulic system, where further specimens were added to the experiment. Therefore, the experiment duration for $\sigma = 18\,\% f_c$ is 7 days shorter in comparison to the remaining specimens. Ultrasonic signals were collected approximately every 10 min using a *Raspberry Pi*-based measuring system [127]. CWI parameters result from evaluation using a fixed reference, immediately after the initial load application. The environmental conditions were 20 °C and 65 % RH. Like the bearing conditions, they stayed unchanged during the entire experiment. The CWI analysis by means of a fixed reference signal, immediately after load application, showed a signal velocity variation corresponding to the characteristic strain evolution during concrete shrinkage and creep. For pure shrinkage, we observed –0.4 % dv/v after half a year, whereas the specimens exposed to external loads showed values up to –3.8 % dv/v for the maximum load. The resulting values of dv/v sequence according to the amount of the applied load. After unloading, all values of dv/v spontaneously returned towards their original values. As all specimens are simultaneously exposed also to shrinkage, the remaining strains cause the CWI parameters not to return completely to zero. However, there are additional changes to the material, regarding strain after unloading. Furthermore, creep recovery of the material appears, also detectable in both strain and velocity variation. Signal correlation coefficients CC significantly decrease instantly after load application as a consequence of large deformations. They are considered ambiguous, especially for small values of CC, i.e., for most creep scenarios. For pure shrinkage, CC remained above 0.85 during the entire experiment, where its development reflects the typical shrinkage behavior, validated by this specimen's strain.

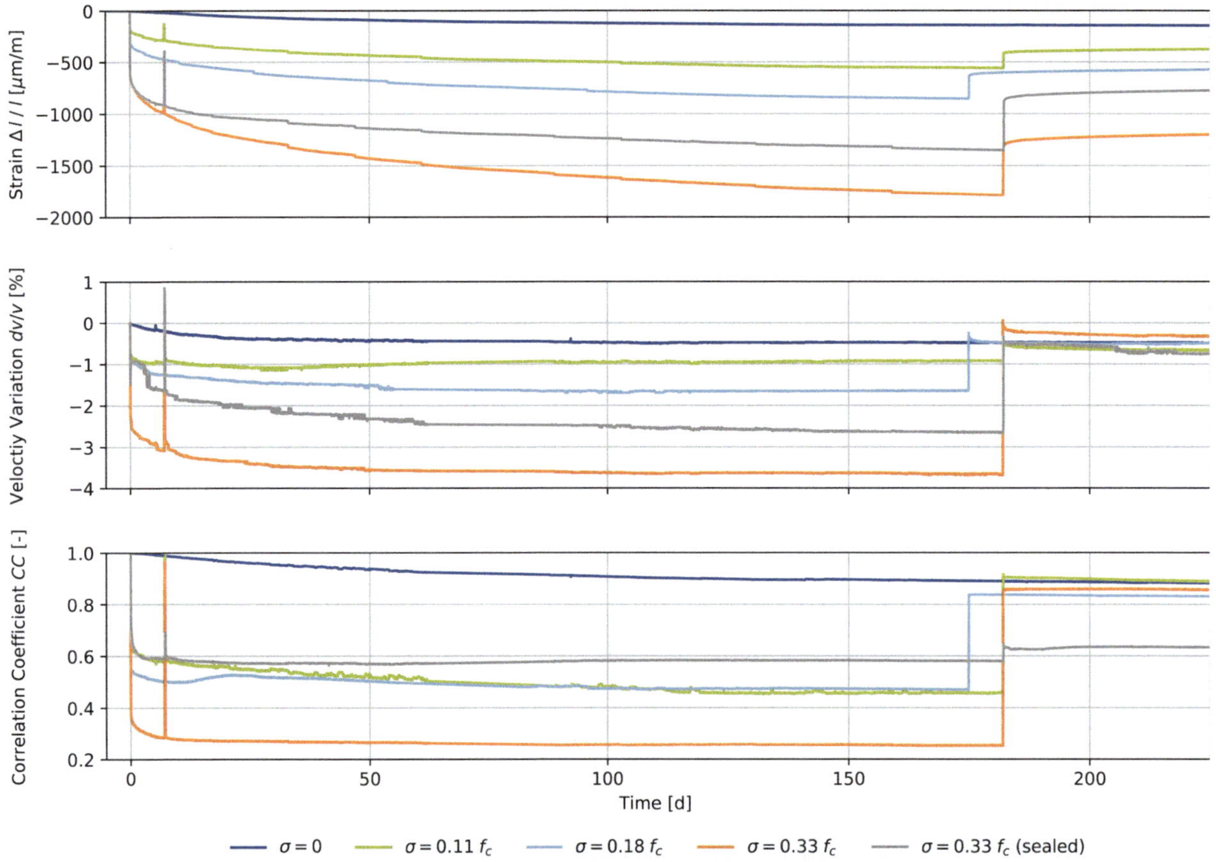

Figure 26: Shrinkage and creep experiment with different constant stresses referring to the setup in Figure 21: strains $\Delta l/l$, velocity variations dv/v and signal correlation coefficient CC over time.

4.5 Discussion

The propagation of Coda waves in concrete is affected by various external impacts that may cause microstructural changes to the material. To interpret the total signal changes, the single effects must be separable, particularly regarding reversible and irreversible changes, or damage. By means of our experiments, we can specify the relations between externally applied mechanical loads, i.e., uniaxial compression and tension, and immediate changes to the ultrasonic signals, i.e., signal correlation and velocity variation, using Coda Wave Interferometry methods. We indicate these relations as calibration curves for a normal concrete, whereas we assess a full range of such calibration curves as key to separate reversible from irreversible changes. This range must be extended by further impacts, i.e., temperature and moisture for the identical material in the future. As we produced a standard concrete using Ordinary Portland Cement, quartzitic aggregates, and a w/c ratio of 0.45, we evaluate the relations as references valid for a broad range of concrete mixes. However, we expect deviations for concrete mixes with varying cement stone and aggregate stiffness due to the modified crack formation process. During the compression experiment, we consider CWI evaluation techniques with a single fixed reference not able to cover the entire stress range between the unloaded state and material failure due to major material changes, especially while irreversible damage advance. The evolution of dv/v can therefore be ambiguous due to abrupt changes to the maximized signal correlation $CC(t,\varepsilon)$, but without a significant physical material change. A fixed reference signal in the unloaded state has proven to be insufficient to detect advancing damage. Therefore, we propose the use of CWI evaluation for a range, where only moderate material changes are expected. We address this hypothesis by the introduction of a tailored method, which switches the reference signal every time the signal correlation drops between a fixed value of 0.7. We recognize further optimization potential regarding this condition to cover small- and large-scale material changes at the same time using condition-based signal reference. Furthermore, we stated a calibration curve for the impact of uniaxial tension on the same concrete mix using one fixed reference signal at the unloaded state. In this experiment, we found nearly linear relations of dv/v and CC for increasing stress as a consequence of a purely uniaxial stress state. Thus, linear crack evolution of single cracks was observed up to failure, completed by the specimens' fracture forming one single macro crack. For both experiments, we showed irreversible changes in the velocity variation dv/v by a remaining velocity decrease for both compression and tension, relating to irreversible micro-cracks in the material. In addition to immediate responses of the CWI parameters, we extended our experiments by investigation of concrete-characteristic shrinkage and creep behavior. These phenomena were monitored in the time domain and are, in contrast to calibration curves, considered to be long-term features. We showed that shrinkage and creep behavior is, in addition to conventional techniques such as strain measurement, representable by evaluation of the velocity variation dv/v. By implication for long-term monitoring processes, it is pivotal to compensate for concrete-typical effects when interpreting two far apart states in time. These time-dependent changes must be considered for signal comparisons by means of calibration curves as the reference may have changed, even if the impacts of stress, temperature and moisture are identical. Further phenomena such as carbonation, freeze attack, or steel corrosion in reinforced concrete will have to be accounted for beyond our investigations.

4.6 Conclusions

Coda Wave Interferometry is a superiorly sensitive method to detect weak changes to heterogeneous materials such as concrete. To establish this technology as an early-warning system for real structures, we analyzed the relations between the correlations and velocity variations of the ultrasonic signals using the stretching technique and externally applied mechanical loads, which we call calibration curves. These curves could serve as the basis for translation of multidimensional fields of velocity variations into stress changes and damage, even for more-than-one-dimensional stress states and on larger specimen scales beyond our investigations. We calculated the relation between velocity variation of the full-length ultrasonic signals for both uniaxial compression and uniaxial tension using a linear regression for elastic material behavior up to 30 % of the respective stress which agrees with previous studies [46, 104, 142]. Furthermore, we evaluated shrinkage and creep processes equally by means of dv/v and CC to indicate long-term changes to the ultrasonic signals. These long-term changes must be considered for signal comparisons over time. Calibration of the effect of mechanical loads adds to the list of all conceivable impacts on concrete structures such as temperature or moisture, and must be quantifiable to separate reversible from irreversible changes, i.e., damage. With our research, we contribute to characterizing the impact of mechanical loads on ultrasound propagating in concrete.

5 Monitoring Early Cement Hydration with Coda Wave Interferometry [3]

Abstract - *The inorganic chemical reaction of hydraulic binders, such as cement with water, is a complex mechanism and a vital part of every building process involving concrete. During this hydration reaction, the system develops strength due to the formation of various solid hydration phases over time. In the case of ordinary Portland cements (OPC), the main hydration phase is called C-S-H phase, and is responsible for the early- and long-term strength development of cement. Therefore, the formation of this phase determines the time for continuing construction work, e.g. removing formwork or concreting of the following structures. Under laboratory conditions and for relatively small samples, methods such as quantitative X-ray diffraction analysis are able to monitor the progress of hydration over time. However, it is more challenging to monitor the hydration of larger structures under non-laboratory conditions.*
In this study, we propose a sensitive and robust monitoring method to detect and quantify the formation of various hydration phases during the early hydration of OPC using Coda Wave Interferometry, based on the evaluation of ultrasonic waves with embedded piezoelectric sensors. We investigated the signal correlation and velocity perturbation of ultrasonic signals during the first 24 hours of OPC hydration. We identified periods during which these parameters varied significantly and periods during which the signals stabilized, in our case approximately 18 hours after initiation of the reaction. Simultaneously, we monitored the hydration of OPC using similar but smaller samples by means of heat flow calorimetry and a Vicat penetration test. We predicted the evolution of the main hydration phases' mass by thermodynamic modeling and validated our results as well as the assignment of characteristic periods to the formation of specific hydration phases in the time domain. Our proposed method adds to the spectrum of ultrasound-based methods for hydration monitoring. Our approach aims at a more sophisticated prediction of the early-age strength development of cement and concrete, even under site conditions and for large structures.

---

[3] This article appeared in its original form in the peer-reviewed Proceedings of the 13th International Workshop on Structural Health Monitoring in Stanford, 2021. Lancaster, PA: DEStech Publications, Inc. [152]. It is adpoted verbatim in this chapter. The authors' work share is listed in Appendix 2.

5.1 Introduction

Today, our built world is composed of a variety of materials. On a global scale, reinforced concrete (RC) is the most widely used structural engineering material [153]. In many cases, the safety and availablity of structures, built from this material, directly relate to the functionality of vital sectors in modern society, e.g. mobility, logistics, and energy. In general, the monitoring of these structures using non-destructive testing methods is a challenge because concrete is a heterogeneous composite material. It comprises several phases that mainly develop during the production of concrete. The chemical reaction of cement with water, or hydration, forms a matrix of hardened cement paste with embedded aggregates, ranging from sand to coarse aggregates, and a system of air pores.

There are a number of promising ultrasound-based techniques for monitoring engineering structures that exploit the reproducibility of ultrasonic (US) signals using embedded sensors. These techniques are cost-efficient given the necessary instrumented volume of the structure. On the other hand, analyses of direct waves or frequency domain analyses often fail because of the great number of heterogeneities, which act as scatterers for the propagating waves. Developing single scatterers, e.g. as a consequence of damage, are not usually detectable when evaluating direct US waves. A more sophisticated technique that exploits the presence of multiple scatterers is called Coda Wave Interferometry (CWI), which was originally developed for geophysical applications. Now, however, current research [119, 34, 26] has started to recognize its potential as an early-warning monitoring system for civil engineering due to its sensitivity to even subtle changes in the material.

5.2 Theoretical Background

5.2.1 Coda Wave Interferometry

The term Coda refers to the strongly scattered tail of an US signal originating from interactions in heterogeneous media [154] in contrast to the evaluation of direct waves. The theory of Coda Wave Interferometry takes advantage of evaluating complex late-onset waveforms by considering them as a summation of the waves that propagate along all possible paths. Although scattering attenuation reduces the signal amplitude for late time windows, the wave information is transformed rather than being cancelled [19] because the waves spend a longer time in the material. Therefore, they sample a specific region within the

medium multiple times. The change in the wavefield is numerically expressed by comparing the unperturbed $u(t')$ and perturbed waveforms $\tilde{u}(t')$ using time-shifted correlation coefficients with a center-time t, and a window length from t_1 to t_2 [18]. The stretching technique [155] interpolates the Coda at time $t(1-\varepsilon)$, where the velocity variation ε of the Coda derives from a veloctiy perturbation dv between two signals.
Here, ε represents the velocity variation, which is the ratio of a veloctiy perturbation dv of a waveform compared to the Coda wave velocity of the reference waveform.

$$\varepsilon = \frac{dv}{v} \tag{5.1}$$

Equation (5.2) compares two US signals using the correlation coefficient CC:

$$CC(\varepsilon) = \frac{\int_{t_1}^{t_2} u(t)\tilde{u}(t(1-\varepsilon))\,dt}{\sqrt{\int_{t_1}^{t_2} u^2(t)\,dt \int_{t_1}^{t_2} \tilde{u}^2(t(1-\varepsilon))\,dt}} \tag{5.2}$$

Equation (5.3) describes the cumulative velocity variation ε_c of the nth waveform:

$$\varepsilon_c = \begin{cases} \varepsilon_n, & n = 1 \\ (\varepsilon_{n-1}+1)(\varepsilon_n+1)-1, & n > 1 \end{cases} \tag{5.3}$$

Coda Waves correspond to multiply scattered wave paths due to numerous heterogeneities. The wavelength is typically smaller than the size of the heterogeneities which are, in turn, smaller than the mean free path. In concrete, these conditions lead to frequencies above 100 kHz, whereas frequencies between 20 kHz and 150 kHz characterize the simple scattering regime [19]. In general, scattering and wave propagation depend on the concrete composition. Microstructural models of concrete can be used to simulate wave propagation for arbitrary compositions [24, 23].

5.2.2 Thermodynamic Modeling of Cement Hydration

Ordinary Portland cement primarily comprises the clinker phases alite (C_3S), belite (C_2S), aluminate (C_3A), and ferrite (C_4AF), where two reaction types dominate the early hydration: the silicate reaction, in which hydration of C_3S results in the formation of C-S-H and CH, and the aluminate reaction in which ettringite is formed [156, 157, 158].
The conventional way of monitoring the in-situ hydration and hardening process of OPC is evaluation of the heat evolution by isothermal heat conduction calorimetry. The heat flow during early hydration reflects the main cement hydra-

tion stages: rapid initial process (I), induction period (II), acceleration phase (III), and retardation period (IV) [158] (Fig. 29), where the acceleration phase and retardation period form the hydration peak.
Thermodynamic modeling of cement hydration, e.g. using Gibbs Energy Minimization Software (GEMS), allows conclusions to be reached about durability-related and mechanical characteristics of cementitious materials over a long-time scale. Methods, available software, and databases are continously improving [159] and state-of-the-art modeling approaches consist of two stages: i) determination of the dissolution rate of the anhydrous cement phases based on experimental data, and ii) prediction of the hydration phase assemblage at given times depending on the predicted dissolution rates of the anhydrous phases and the data of cement-specific phases [160]. Several approaches for predicting the dissolution of anhydrous cement phases exist [159], such as the model presented by Parrot & Killoh [161]. The concept of Gibbs free energy minimization [162, 163] is the basic principle behind models for predicting the formation of hydration phases. However, this approach necessarily assumes thermodynamic equilibrium, even though the solid and liquid phase are not in equlibirium, especially during early-age hydration.

5.3 Methods

5.3.1 Experimental Setup and Signal Data Processing

We produced a concrete mix using OPC CEM I 42.5 R and a water-to-cement-ratio of 0.45. The cement's density was 3.12 g/cm^3 and a Blaine test yielded a specific surface of 3400 cm^2/g. The aggregates were crushed stones with predominantly quartzitic components with an approximated A/B 16 grading curve according to the German standard *DIN 1045-2*. Additionally, 2.42 wt.-% cement of a polycarboxylatether-based superplasticizer, added to the mixing water, improved the concrete workability and the sensor-to-concrete coupling condition. Fig. 27 (a) shows the two embedded US transducers (ACSYS S0807) with a center frequency of 60 kHz, (b) the experimental setup, and (c) one specimen after completion of the reaction. The sensors were fixed using pipes which are still visible after removing the formwork. The portable and low-cost measurement device was specifically designed for concrete monitoring [127]. The system transmits pulses at a frequency of 60 kHz to one selectable transducer and records the resulting signal with a sampling rate of 1 MHz and 12,000 samples with a resolution of 14 bits. We subsequently compared the signals without further averaging. We pre-processed the US signals by applying a second-order

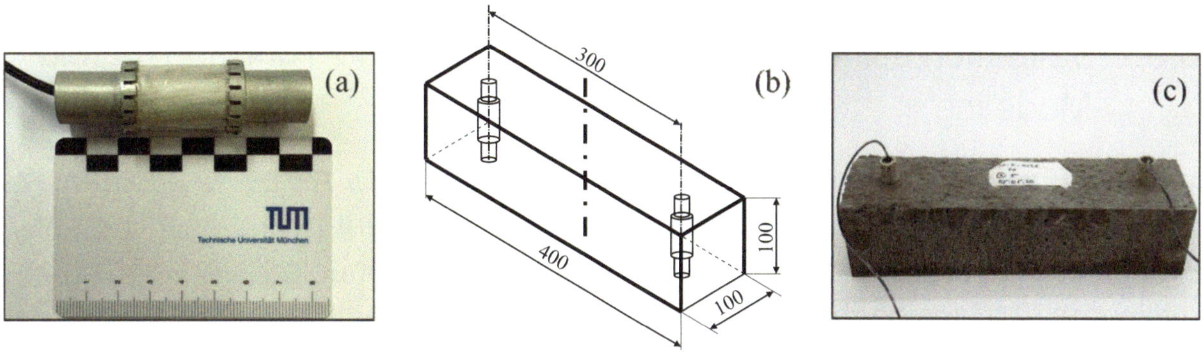

Figure 27: (a) Cylindric piezoelectrical US sensor with a center frequency of 60 kHz. (b) Sketch of experimental setup. The distance between the two embedded sensors is 300 mm. (c) Specimen with dimensions of 400 x 100 x 100 mm^3, 24 h after production.

Butterworth bandpass filter in the time domain with cutoff frequencies of 1 kHz and 120 kHz without tapering. The average amplitude of the first 50 samples determined the offset compensation for each signal. To eliminate electromagnetic crosstalk between the transducers, the first 100 samples of the signal were set to zero. The entire hydration process was monitored by computing the cross-correlation coefficient and the velocity perturbation from each pre-processed signal, collected every 7 minutes. These parameters were calculated using a time window length of 2,500 samples which equates to 2.5 ms, and a stepwise signal comparison. Fig. 28 shows four US signals after pre-processing at different times during the hydration process with rising amplitudes and axes scales.

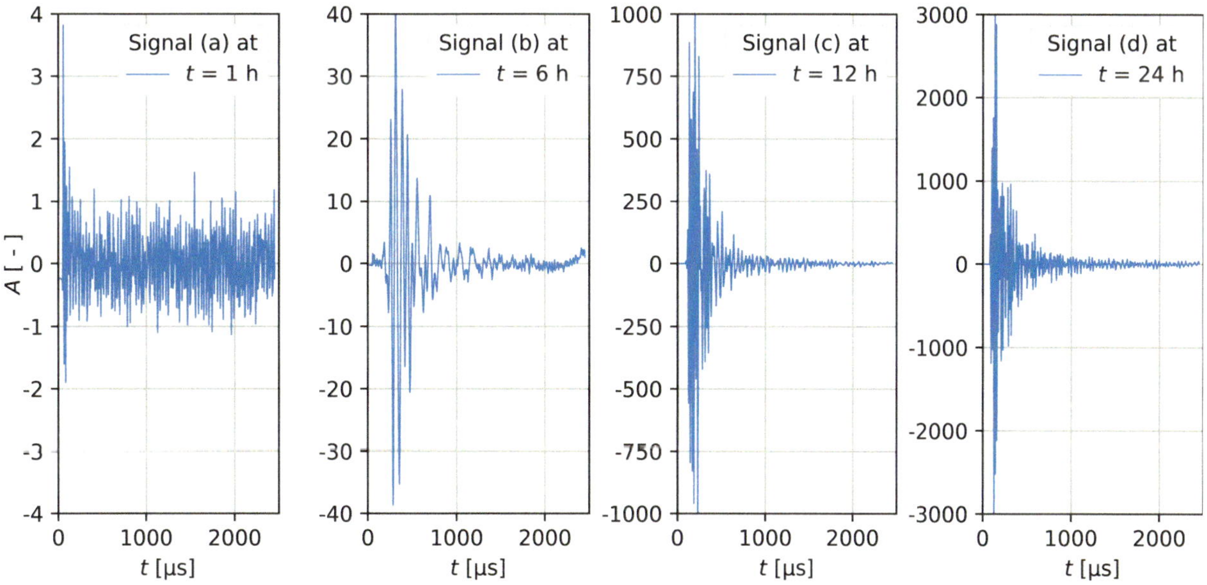

Figure 28: Pre-processed US signal at different times during OPC hydration, referencing water addition. The sampling rate was 1 MHz for 2500 samples.

5.3.2 Validation using Heat Flow Calorimetry, and Thermodynamic Modeling

We used an eight-channel conduction calorimeter TAM Air by TA Instruments to measure the heat flow during the first 24 h of cement hydration. The anhydrous OPC was weighed into 20 ml glass ampoules to enable water addition and mixing at 20 °C inside the calorimeter. The resulting heat flow curve represents the average of three simultaneous measurements. In addition, we used a standardized Vicat needle with a 1 mm² circular cross section to monitor the OPC hardening process by means of penetration depth over time according to standard *DIN EN 196-3*. We used GEMS [163] to predict the dissolution of the four main clinker minerals in the anhydrous cement by means of a modified Parrot & Killoh model [164, 161] taking into account the composition of the anhydrous cement as determined by quantitative X-ray diffraction analysis, its specific surface area determined by the Blaine method and the curing temperature. The formation of the following phases was neglected: AFt solid solutions containing iron, carbonate or hydroxy end members AFt-AlFe-ss, AFt-s-c-ss, AFm-s-OH-ss. Furthermore, hydrogarnet (C3AH6), siliceous hydrogarnet (C3FS0.84H4.32 and C3FS1.34H3.32), C4AcH9, M-S-H, gibbsite, thaumasite, brucite, dolomite, magnesite, titanite, hematite, goethite, pyrite, syngenite were not considered because their formation is kinetically hindered at the prevailing temperature and pressure.

5.4 Results

After preprocessing, we evaluated the US signals using CWI according to Equation (5.2) and Equation (5.3). As the concrete was processed first, we show the stepwise correlation coefficient CC, velocity perturbation dv/v, and cumulative velocity perturbation ε_c between 1 h and 24 h after water addition in Fig. 29. During the first 6 h of hydration, CC and dv/v have unstable values due to a small signal-to-noise ratio. However, these parameters converge between (b) and (c) for CC, and between (b) and (d) for dv/v and ε_c. The comparison between both CWI parameters reveals a more sensitive response of the velocity variation to phase changes in the materials during hydration, because it is still detecting changes in the material phase assemblage after 12 h.

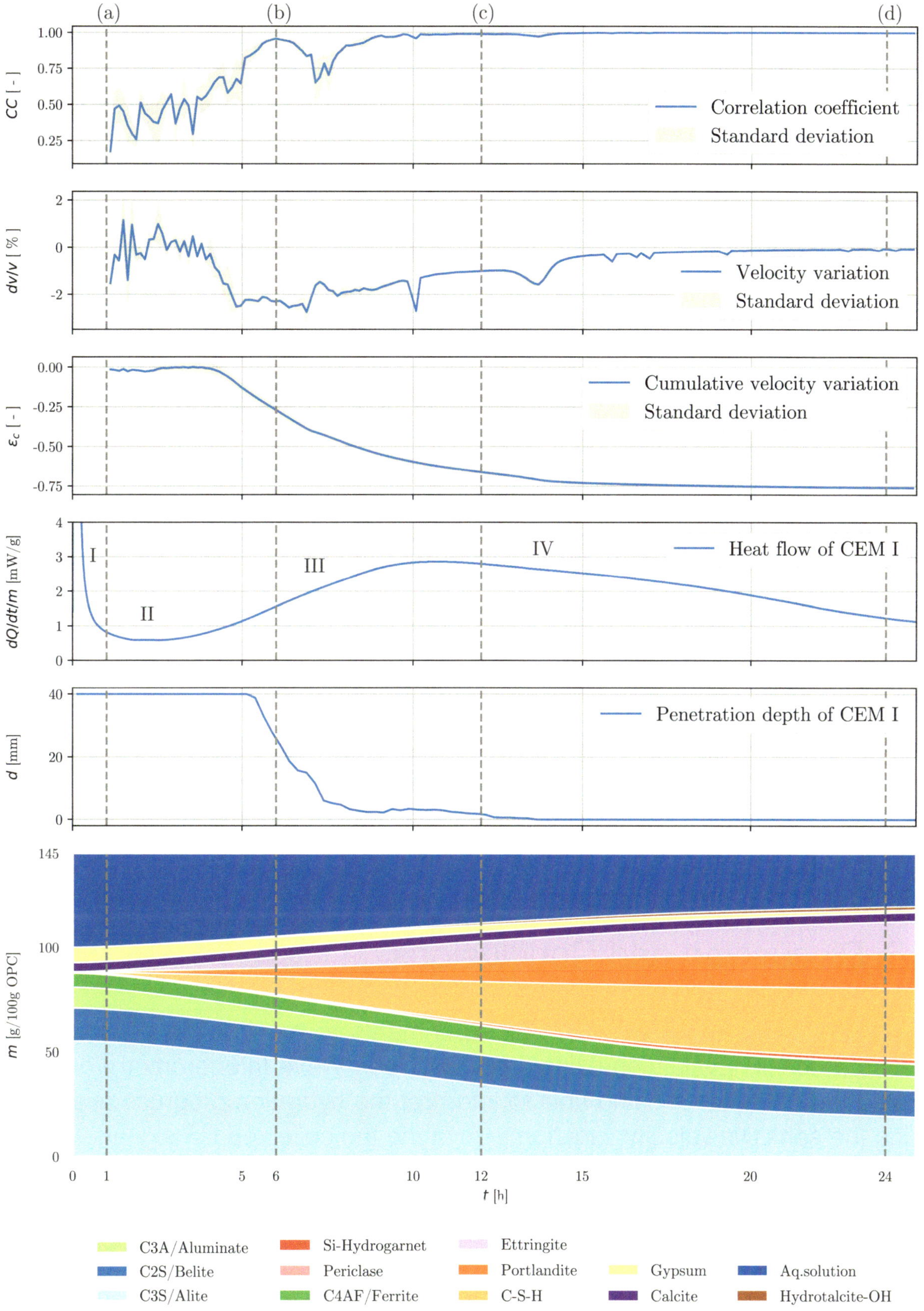

Figure 29: Correlation coefficient CC, velocity variation dv/v and cumulative sum ε_c using CWI with a stepwise reference. Heat flow dQ/dt *with characteristic periods [158], and penetration depth* d*, all averaged from three specimens of CEM I. Assemblage of phase masses m per 100 g OPC in the cement paste, predicted by thermodynamic modeling over hydration time t. Positions (a) to (d) refer to the US signals from Fig. 29.*

For validation purposes, we show the experimentally obtained isothermal calorimetric heat flow at 20 °C with the initial peak immediately after water addition and the following induction period up to approximately 3.5 h of hydration. The acceleration period reaches a heat flow maximum at approximately 11 h. During the following retardation period, the heat flow continuously decreases and approaches the x-axis asymptotically. Simultaneously, the Vicat test shows the major decrease in the penetration depth around 6 h upon entering the acceleration phase. In order to identify the individual mechanisms of cement hydration, we compared the experimental results with the modeled evolution of the phase assemblage (Fig. 29, bottom). Alite dissolution starts approximately at the hydration heat flow peak, where the point of inflection of the dissolution reaction correlates with the peak of the heat flow. Furthermore, the dissolution of the main clinker phases, especially C_3S, correlates with the obtained heat flow curve.

The comparison of the cumulative velocity variation ε_c with the heat flow shows a consistent identification of the transition between the induction period (II) and the acceleration phase (III) after approximately 4 h. The GEMS model confirms this conformity as the dissolution of the main clinker phases also accelerates after 4h. In addition, ε_c and the simulated dissolution show a similar development and concurring points of inflection after 8 h of hydration. ε_c convergence occurs after about 18 h, whereas the simulated dissolution reactions mainly decelerate after 20 h. Both monitoring of the velocity variation and the phase assemblage model show convergence within the retardation period (IV) and significantly after having reached the heat flow peak.

5.5 Discussion and Conclusions

In our study, we conducted several experiments to monitor the concrete hydration process. Evaluation using stepwise Coda Wave Interferometry demonstrated the potential of being able to interpret the hydration progress over time using the sensitive velocity variation dv/v of the the collected US signals.

We were able to show the correlation between the CWI method and conventional techniques, i.e., heat flow calorimetry and Vicat test, for a characteristic hydration reaction of Ordinary Portland Cement. Furthermore, we demonstrate conformity of the velocity variation development with the dissolution of the main clinker phases by modeling the early-age phase formation process using CemGEMS.

Generally, Coda wave interferometry is applied for monitoring slight changes in the material. However, we are also able to use stepwise CWI to monitor time-dependent processes, i.e., material phase changes during hydration, by interpreting the dv/v development as a characteristic feature. Also, our technique is insensitive to the actual starting time, or knowledge of the exact source-receiver position of the US sensors because moderate modifications have no effect on the velocity change estimation. We consider these conditions advantageous for the technology's application under actual conditions when estimating the progress of the hydation process, and therefore, the early-age strength development of concrete structures.

6 Conclusions

6.1 Summarized Impacts on the Velocity Variation

Coda wave interferometry is a highly sensitive method to detect even weak changes in heterogeneous media like concrete using the velocity change of elastic waves propagating in the material. The ultrasonic velocity variation can represent various impacts on civil engineering structures, many of which have been discussed in literature (Chapter 2.3.2).

We can collect one ultrasonic signal at a time for each sender-receiver pair in a structure and determine the total velocity change between the measurements at a reference state and the evaluated state. The measured velocity variation can be expressed as a sum of three terms that change at different rates over time as in Equation (6.1).

The transient term covers the impacts from state changes due to temperature, stress, or moisture, where the changes in dv/v are expected immediatly, e.g., for stress, or over a few hours up to days, e.g., for temperature and moisture in large structures. A second stationary term includes velocity variations that steadily develop over longer periods, i.e., in the range of over months or years. This term is affected by long-term strain or expectable chemical changes of the phase composition.

The first two terms are considered to be reversible, whereas a third term includes changes that are induced by any irreversible material change. Therefore, any value greater than zero is an indicator of damage in the structure.

$$\left(\frac{dv}{v}\right)_{measured} = \left(\frac{dv}{v}\right)_{transient} + \left(\frac{dv}{v}\right)_{stationary} + \left(\frac{dv}{v}\right)_{irreversible} \tag{6.1}$$

Figure 30 presents the individual effects on Equation (6.1) coming from the structural system in greater detail . The transient impacts on the ultrasonic velocity change from temperature, stress, and moisture have been investigated in the articles, here in Chapter 3 and Chapter 4. The extreme case of phase changes during early cement hydration has been subject of the article in Chapter 5.

The left side of Equation (6.1) is the result of ultrasonic measurements, followed by the calculation of dv/v, while the irreversible term on the right side is usually the critical one and should be determined to identify any damage in the structure. To solve the equation, both the remaining terms must be evaluated in detail. Equation (6.2) further breaks down the transient term of dv/v, where individual effects of temperature T, saturation degree S of the material, i.e., the

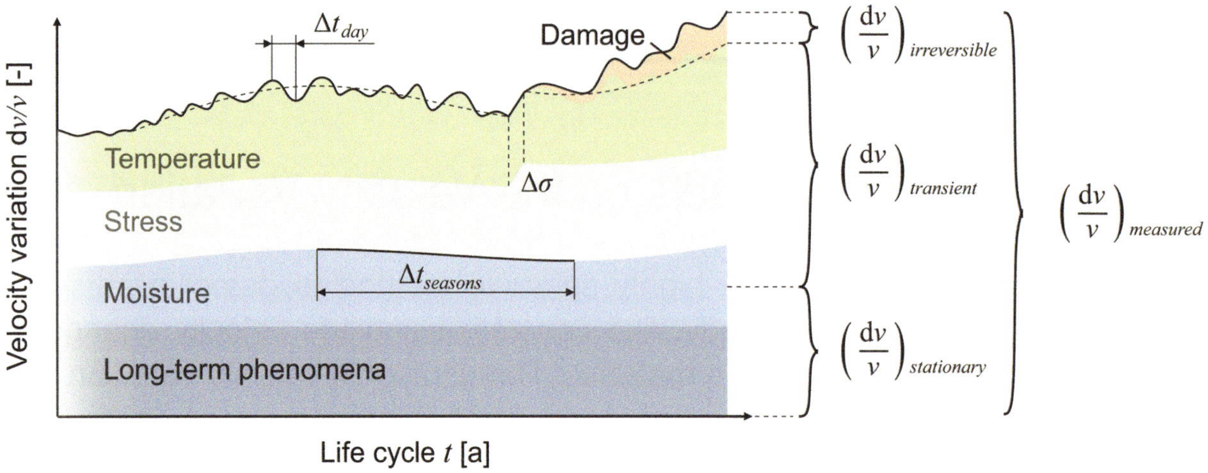

Figure 30: Summarized effects on the velocity variation dv/v refined from Figure 3. Originating from the measured total velocity variation, dv/v splits into three terms according to Equation (6.1). Transient changes occur on a short time scale, whereas stationary changes are only visible over a long period. Assuming that these two terms can be predicted, any remainder can be considered an irreversible change that indicates damage in the structure.

pore space, and stress σ are included. In this notation, x_0 refers to the value of each variable at the reference state, while Δx refers to its change.

$$\left(\frac{dv}{v}\right)_{transient} = \frac{dv}{v}(T_0, \Delta T) + \frac{dv}{v}(S_0, \Delta S) + \frac{dv}{v}(\sigma_0, \Delta\sigma) \tag{6.2}$$

Both **temperature** $(T_0, \Delta T)$ and **saturation degree** $(S_0, \Delta S)$ are variables that must be considered together, as they affect each other due to the temperature-affected moisture content of the structure's environment and the moisture gradient arising in the material. It is important to note that the saturation degree of the pore space affects the anelastic seismic properties of concrete, in which context we demonstrated an increasing attenuation for higher saturation using the seismic quality factor Q. However, S is usually not directly measurable and must be expressed by external variables, e.g., relative humidity RH or the electrical resistance of the electrolyte Ω [165]. Equation (6.2) emphasizes the impact on dv/v as a result of the state change of the material due to the saturation degree rather than using an external variable, for which a delayed system response should be considered.

The relationship between temperature and dv/v is quasi-linear for the same water-to-cement ratio, which we found to be comparable with previous studies to within an order of magnitude [107, 137]. Regarding relative humidity variation, we discovered a steady decrease of dv/v for an increase of water in the pore system between 35 % RH up to 80 % RH at a reference temperature of 20 °C. Furthermore, a saturation degree above this value can disrupt the trend, where

dv/v increases compared to the lower values of RH. However, this observation is associated with a high degree of uncertainty due to attenuation effects, as demonstrated by Q, and a low cross correlation value CC between the ultrasonic signals. Combining all observations and the additional consideration of the water-to-cement ratio, which influences the pore space, leads to a third order polynomial model that covers the variation of the three variables and their effect on the velocity variation dv/v at a high degree of confidence.

The saturation degree is especially indicative of the fact that any relationship stated for the individual arguments depends on the material composition because the amount of water in the pore space is also affected by the volume of the pore space, which again depends on the water-to-cement ratio w/c. We demonstrated this in the polynomial model, which had dominant terms depending on w/c. Furthermore, also elastic, anelastic, and scattering properties may also change for different material mixes. This implies that any experimentally determined relationship regarding both the transient and the stationary terms is only valid for the particular material. The relationships between the transient and stationary impacts on the velocity variation can either be experimentally determined for the particular case, for which Equation (6.1) should be solved, or predicted by material models in combination with wave propagation simulations. Such models can be calibrated by means of the physicial relationships and the data from our research, with the objective of being transferable to arbitrary material mixes. Nevertheless, in order to compensate for velocity changes due to temperature and the saturation degree, T and S of the structure must be known and measured continuously, e.g., in the same grid as ultrasonic transducers are embedded.

The relationship between **stress** $(\sigma_0, \Delta\sigma)$ and the velocity variation is also affected by the material mix as the crack formation process changes for different material mixes. We find that CWI using a single reference is not able to cover the entire stress range between the unloaded state and material failure and propose a method that switches the reference signal every time the signal correlation drops between a fixed minimum value of the cross correlation CC between two ultrasonic signals. This approach holds especially true for the compression case since the crack formation process is more complex than uniaxial tension, during which mostly individual cracks develop. We demonstrated that a stress increase can be approximated by a linear relationship in the elastic range up to 30 % of the material strength for both compression and tension, whereas an irreversible velocity decrease occurs above this load level.

The study (Chapter 4) explicitly shows that the influence of the cross correlation value CC on the calculation of dv/v must be considered and its computation must be standardized during the entire evaluation cycle. This becomes especially important for major changes in the structural system, i.e., irreversible ones or long-term changes with a great effect on the ultrasonic waveforms. Equation (6.3) provides the arguments of the long-term stationary term, where the effects of strain ε and chemical phase changes Ψ are included.

$$\left(\frac{dv}{v}\right)_{stationary} = \frac{dv}{v}(\varepsilon_0, \Delta\varepsilon) + \frac{dv}{v}(\Psi_0, \Delta\Psi) \tag{6.3}$$

Typicial phenomena that induce long-term **strain** $(\varepsilon_0, \Delta\varepsilon)$ in concrete are shrinkage and creep, in which the resulting strains have an impact on the velocity variation in addition to the immediate change of dv/v due to stress. We showed that the characteristic shrinkage and creep behavior of concrete can be measured for several load levels using CWI. Consequently, these changes must be considered for long-term state comparisons using CWI because these changes appear in the stationary term.

Furthermore, the velocity variation is affected by any **chemical phase changes** $(\Psi_0, \Delta\Psi)$. Phase changes are a continuous process in cementitious materials, whereas two types of phase changes should be distinguished in this context. First, early hydration is the most extreme case, during which cement and water react, create strength, and develop into the solid form cement paste. We demonstrated that the velocity variation development using ultrasonic signals compared in steps correlates with the dissolution of the main clinker phases over time. CWI exhibits a similarly high level of accuracy in detecting characteristic hydration periods compared to conventional techniques, e.g., heat flow calorimetry and the Vicat test, but also compared to a thermodynamic model of early-age phase formation using CemGEMS.

Although we demonstrated that ultrasonic velocity changes can detect the early-age strength development of concrete during the early hydration, CWI is usually designed to detect weak changes in a structural concrete system, as expressed by Equation (6.1). As a result, phase changes addressed in Equation 6.3 are rather long-term and must be considered as a source for velocity changes.

Separating the individual arguments for the transient and the stationary terms is not always possible, nor is doing so always necessary because some effects are inseparably linked. By way of example, we can consider the separation of the impacts of drying shrinkage-induced strain on dv/v, and the impact of saturation S, induced by a varying external relative humidity, which always occur simultaneously. Another example of an inconclusive separation would be the response of S at different locations in the structure. The almost immediate response in the edge zone can be addressed using the transient term, while the diffusion-driven water transport in deeper sections of the structure is compara-

tively slow and can be addressed using the stationary term (Figure 30).
In order to make Equation (6.1) fully interpretable, we will need material models that are able to cover all of the effects summarized for individual material mixes with arbitrary scattering properties. In turn, the scattering properties should be defined via material parameters, e.g., elasticity, pore and aggregate size, which can be easily determined for each concrete, in which case both physics-based and data-driven surrogate models are conceivable. As soon as the individual arguments for the transient and the stationary terms can be predicted by means of concrete models, the total measured velocity variation can then be evaluated with respect to any remaining value that can be associated with the irreversible term as an indicator for damage.
However, this requirement only exists when a full interpretation becomes necessary. This is mostly the case regarding the identification of long-term damage development together with the separation from other long-term phenomena that are non-destructive. Given its sensitivity to microstructural changes, CWI is a suitable technology for the identification of spontaneous or short-term damage development in general, even if the transient and the stationary terms or the scattering properties are not fully known. In this case, even without an exact interpretation of the causes for velocity changes, CWI is still applicable as an early-warning system.

6.2 Coda Wave Interferometry for Early-warning Systems

Given that they are collected using CWI, time-series data are generally more complex to interpret than evaluating scalar differences, upon which most decisions regarding maintenance and repair are currently based in traditional engineering. On the one hand, time-series data pose a special challenge because new concepts for fast and reliable decision making must be established. As these decisions are often safety-critical and include far-reaching consequences. For example, when making decisions about the future of important and expensive infrastructure, building a foundation of trust in such structured data is a long-term process, and the interlocking data, methods, and processes must be brought into a sophisticated and robust framework. Furthermore, no unique solution or ideal monitoring system exist, and the composed technologies and models must be assessed individually for each building.
On the other hand, a large number of buildings exist worldwide, whose condition is unmonitored and therefore not fully known. Not only existing buildings are suitable, but new buildings can also be instrumented, whereby CWI is able

to further support the construction process, e.g., track the hydation and strength development. Especially building types, which are cost-intensive to inspect, maintain and repair can profit from novel and innovative methods for condition assessment, e.g., early damage detection. Two characteristic and representative structures in Germany, the *Gänstorbrücke* bridge between Ulm and Neu-Ulm and the *Scheidplatz* underground station in Munich have been instrumented with CWI technology during the project and will be monitored during the upcoming years (Figure 31). For the bridge, the major objective is the prolongation of its service life after the identification of existing damage, using a set of NDT methods including CWI. For the underground station, potential microstructural changes will be investigated in the ceiling during planned static load redistributions. These demonstration sites will proof the potential of CWI on the structural level for buildings in operation where the results from the laboratory scale will be transferred to the structural scale. This in particular includes the separability of different effects on the ultrasonic velocity change, the robustness of the embedded transducers together with the measurement system, and the geometrical localization of any structural change.

Figure 31: Photographs of the two structures instrumented with CWI technology: the Gänstorbrücke *bridge between Ulm and Neu-Ulm instrumented in September 2020 [166] and the* Scheidplatz *underground station in Munich instrumented in June 2022.*

Comprehensive monitoring systems will be part of our future built environment. Such systems will include sophisticated sensor networks in connection with the matching NDT methods, data processing techniques, and interfaces that together support the decision making of engineers and authorities in the framework of a digital twin. Even when listing the single system components, it is noticeable that these systems will be characterized by a high degree of complexity and cannot be comprehensively controlled by a single person. Experts from multidisciplinary fields will have to continue working together on reliable, robust, and interpretable solutions, thus ensuring that systems remain controllable and that the final decision on suitable measures is left entirely to humans.

Appendices

1 Note on Related Literature

Publications

F. Diewald, N. Klein, C. M. Hechtl, T. Kraenkel, C. Gehlen, "Efficient Labelling of Air Voids in Hardened Concrete for Neural Network Applications Using Fused Image Data", Proceedings of the 3rd RILEM Spring Convention and Conference (RSCC 2020), 2021.

V. Holla, G. Vu, J. J. Timothy, F. Diewald, C. Gehlen, G. Meschke, "Computational Generation of Virtual Concrete Mesostructures", materials, 14(14), 3782, 2021.

G. Vu, F. Diewald, J. J. Timothy, C. Gehlen, G. Meschke, "Reduced Order Multiscale Simulation of Diffuse Damage in Concrete", materials, 14(14), 3830, 2021.

F. Malm, F. Diewald, K. Pinkert, "Non-destructive Inspection and Monitoring of Fractures in Concrete with Self-healing Properties", Fifth International Conference on Sustainable Construction Materials and Technologies, London, 2019.

Supervised Theses

E. Jägle, "Monitoring Cement and Concrete Hydration Using Coda Wave Interferometry", *Master's Thesis*, Technical University of Munich, 2022.

H. Olabi, "Charakterisierung der Mikrorissbildung von Beton", *Bachelor's Thesis*, Technical University of Munich, 2022.

J. Ernstberger, "Charakterisierung der Mikrorissbildung in Beton in Folge zentrischer Zugspannung", *Bachelor's Thesis*, Technical University of Munich, 2021.

F. Niewöhner, "Mikrorissschädigung von Betonproben infolge von Kriechen bei unterschiedlichen Druckbelastungen", *Bachelor's Thesis*, Technical University of Munich, 2021.

M. Härder, "Untersuchungen zum Feuchteverhalten tragender Lehmbaustoffe", *Master's Thesis*, Technical University of Munich & Bundesanstalt für Material-
have read and agreed to the published version of the manuscript.

A. Kamberi, "Monitoring der Entwicklung der Frühfestigkeit von Beton mittels diffusem Ultraschall", *Bachelor's Thesis*, Technical University of Munich, 2021.

S. Rier, "Untersuchungen zum Einfluss des Kriech- und Schwindverhaltens von Festbeton auf diffusen Ultraschall mittels der Codawellen-Interferometrie", *Master's Thesis*, Technical University of Munich, 2021.

C. M. Hechtl, "Automatisierte Charakterisierung von Luftporen in Festbeton", *Master's Thesis*, Technical University of Munich, 2020.

M. Schwarzer, "Einfluss der Betonzusammensetzung auf das Codasignal", *Master's Thesis*, Technical University of Munich, 2020.

M. Roger, "Detektion von Spannstahlbrüchen mittels Acoustic Emission - Durchführung und Auswertung von Bauteil- und Bauwerksversuchen", *Master's Thesis*, Technical University of Munich, 2020.

2 Author Contributions

The structure of the author contributions is in accordance with the author contributions structure of the journal, in which the original article appeared first.

Chapter 3 – Impact of Temperature and Moisture Variations on Coda Waves in Concrete

Conceptualization, Fabian Diewald; Data curation, Fabian Diewald; Formal analysis, Fabian Diewald; Funding acquisition, Marine, Denolle, Jithender Timothy and Christoph Gehlen; Investigation, Fabian Diewald, Marine Denolle and Jithender Timothy; Methodology, Fabian Diewald, Marine Denolle and Jithender Timothy; Project administration, Fabian Diewald and Christoph Gehlen; Resources, Marine Denolle and Christoph Gehlen; Software, Fabian Diewald; Supervision, Marine Denolle and Christoph Gehlen; Validation, Fabian Diewald, Marine Denolle, Jithender Timothy and Christoph Gehlen; Visualization, Fabian Diewald; Roles/Writing - original draft, Fabian Diewald; Writing - review & editing, Fabian Diewald, Marine Denolle, Jithender Timothy and Christoph Gehlen.

Chapter 4 – Impact of External Mechanical Loads on Coda Waves in Concrete

Conceptualization, Fabian Diewald; methodology, Fabian Diewald; software, Fabian Diewald and Niklas Epple; validation, Fabian Diewald, Niklas Epple, Christoph Gehlen and Ernst Niederleithinger; formal analysis, Fabian Diewald and Niklas Epple; investigation, Fabian Diewald and Niklas Epple; resources, Fabian Diewald; data curation, Fabian Diewald; writing—original draft preparation, Fabian Diewald and Niklas Epple; writing—review and editing, Fabian Diewald, Niklas Epple, Thomas Kränkel, Christoph Gehlen, and Ernst Niederleithinger; visualization, Fabian Diewald; supervision, Thomas Kränkel, Christoph Gehlen and Ernst Niederleithinger; project administration, Fabian Diewald, Niklas Epple, Christoph Gehlen, and Ernst Niederleithinger; funding acquisition, Fabian Diewald, Niklas Epple, Christoph Gehlen, and Ernst Niederleithinger. All authors have read and agreed to the published version of the manuscript.

Chapter 5 – Monitoring Early Cement Hydration with Coda Wave Interferometry

Conceptualization, Fabian Diewald; Data curation, Fabian Diewald and Linda Irbe; Formal analysis, Fabian Diewald and Linda Irbe; Funding acquisition, Christoph Gehlen; Investigation, Fabian Diewald, Linda Irbe, Thomas Kränkel, Alisa Machner and Christoph Gehlen; Methodology, Fabian Diewald; Project administration, Alisa Machner and Christoph Gehlen; Resources, Alisa Machner and Christoph Gehlen; Software, Fabian Diewald and Linda Irbe; Supervision, Alisa Machner and Christoph Gehlen; Validation, Fabian Diewald, Linda Irbe, Thomas Kränkel, Alisa Machner and Christoph Gehlen; Visualization, Fabian Diewald; Roles/Writing - original draft, Fabian Diewald; Writing - review & editing, Fabian Diewald, Linda Irbe, Thomas Kränkel, Alisa Machner and Christoph Gehlen

3 ChatGPT about Future Monitoring Systems for Civil Engineering Structures

The interview was conducted with the chatbot prototype *ChatGPT* [167], launched by *OpenAI* on November 30, 2022. Partly meant seriously.

January 13, 2023
Fabian Diewald: Welcome ChatGPT, you have been launched less than two months ago and attracted quite a lot of attention since then. Everyone is looking forward to your future work in general and in this interview, about your professional assessment on the future of monitoring systems for civil engineering structures!
ChatGPT: Thank you for your kind words. I am continually learning and updating my knowledge, so my understanding of the future of monitoring systems for civil engineering structures may change over time. However, currently, I can tell you that monitoring systems for civil engineering structures are becoming increasingly important as the need for more efficient and cost-effective ways to maintain and repair structures grows. Advancements in technology, such as the use of sensors, drones, and machine learning, are leading to more accurate and efficient monitoring systems. In the future, it is likely that these systems will become even more sophisticated and integrated into the overall design and construction process. Additionally, the use of cloud-based systems and the Internet of Things (IoT) will enable real-time monitoring and analysis of structures, allowing for more proactive maintenance and repairs. Overall, the future of monitoring systems for civil engineering structures is promising and will continue to play a crucial role in the safety and longevity of our built environment.
Fabian Diewald: That was quite a lot for the beginning. Can you elaborate a bit more on how future monitoring systems for integrity assessment will look like in twenty years?
ChatGPT: Sure, in twenty years it is likely that monitoring systems for integrity assessment of civil engineering structures will be highly advanced and integrated. The use of sensors, drones, and machine learning will become more prevalent, allowing for real-time monitoring and analysis of structures. This will enable more proactive maintenance and repairs, reducing the need for costly and disruptive interventions.
One aspect that is likely to be more developed is the integration of monitoring systems with the design and construction process. For example, the use of Building Information Modelling (BIM) will allow for more accurate and efficient monitoring of structures throughout their life cycle.
Another key aspect that will be more developed is the integration of monitoring

systems with the Internet of Things (IoT) and cloud computing. This will enable real-time data transfer and analysis, allowing for more accurate and efficient monitoring and maintenance of structures.
Virtual and Augmented Reality technologies will also be more prevalent, allowing for more immersive and interactive monitoring and analysis of structures. This will enable engineers and maintenance personnel to more effectively visualize and understand the condition of structures, allowing for more accurate and efficient decision-making. Overall, in twenty years, monitoring systems for integrity assessment will be highly advanced, integrated and will enable real-time data collection, analysis, and visualization, allowing for more accurate, efficient, and cost-effective maintenance of civil engineering structures.
Fabian Diewald: What trends do you see for detecting structural damage?
ChatGPT: There are several trends in the development of methods for detecting structural damage that I see as likely to continue in the future:

1. Non-destructive testing (NDT) methods: NDT methods, such as ultrasound, thermal imaging, and X-ray imaging, are becoming increasingly popular for detecting structural damage. These methods are non-invasive, meaning they do not cause any damage to the structure being tested.

2. Machine learning and artificial intelligence: Machine learning and AI techniques are becoming increasingly prevalent in the development of damage detection methods. These techniques can be used to analyze sensor data and identify patterns that indicate structural damage.

3. Remote sensing: The use of drones and other remote sensing technologies is becoming more popular for damage detection. These technologies can be used to survey structures and gather data on their condition.

4. Vibration-based methods: Vibration-based methods, such as modal analysis and operational modal analysis, are becoming more popular for detecting structural damage. These methods can be used to identify changes in the dynamic behavior of a structure, which can indicate damage.

5. Structural health monitoring: This is a process of monitoring the structural integrity of a structure over time, using sensors to detect changes in the structure's behavior, such as cracks or deformations, that might indicate damage.

Overall, the trend is towards the integration of multiple methods, using sensor data and AI techniques to detect the structural damage, and using remote sensing and structural health monitoring for long-term monitoring of the structural integrity.

Fabian Diewald: Will Coda Wave Interferometry be part of such a comprehensive system?
ChatGPT: Coda Wave Interferometry (CWI) is a technique that can be used for the detection of structural damage. The technique uses the seismic coda, which is the residual noise that is left after the main energy from an earthquake has dissipated, to detect changes in the subsurface of a structure.
CWI has been used to detect damage in a variety of structures, including bridges, buildings, and dams, and it has been shown to be effective in detecting damage that is not visible on the surface.
It is likely that CWI will be part of a comprehensive monitoring system for civil engineering structures in the future. The technique can be used in combination with other methods, such as non-destructive testing and vibration-based methods, to provide a more comprehensive understanding of the condition of a structure.
Additionally, the integration of CWI with other technologies such as AI and machine learning, can improve the accuracy of the damage detection and make it more efficient.
It's worth noting that CWI can also be used to monitor underground structures such as mines, oil and gas pipelines, and tunnels. It is a versatile technique that can be used in a wide range of civil engineering structures.
Fabian Diewald: Thank you very much for this overview, ChatGPT! We will see, if your assessment will come true one day.
ChatGPT: You're welcome! It's always a pleasure to share my knowledge with you. It's important to note that my understanding of the future of monitoring systems for civil engineering structures is based on current trends and information available, it's not a definite prediction. However, I believe that the integration of advanced technologies such as AI, machine learning, sensors, drones, and the Internet of Things will lead to more accurate, efficient and cost-effective monitoring systems for civil engineering structures in the future.

References

[1] L. Beltrametti, A. Bottasso, M. Conti, C. Ferrari, and M. Piana, "Effetti della caduta del Ponte Morandi sull'economia," *Rivista di economia e politica dei trasporti*, vol. 2, 2018.

[2] S. Invernizzi, F. Montagnoli, and A. Carpinteri, "Fatigue assessment of the collapsed XXth century cable-stayed polcevera bridge in genoa," *Procedia Structural Integrity*, vol. 18, pp. 237–244, 1 2019.

[3] M. Morgese, F. Ansari, M. Domaneschi, and G. P. Cimellaro, "Post-collapse analysis of Morandi's Polcevera viaduct in Genoa Italy," *Journal of Civil Structural Health Monitoring*, vol. 10, pp. 69–85, 2 2020.

[4] J. Rymsza, "Causes of the collapse of the Polcevera Viaduct in Genoa, Italy," *Applied Sciences 2021*, vol. 11, p. 8098, 8 2021.

[5] P. Clemente, "Monitoring and evaluation of bridges: lessons from the Polcevera Viaduct collapse in Italy," *Journal of Civil Structural Health Monitoring*, vol. 10, pp. 177–182, 4 2020.

[6] C. Gehlen, *Probabilistische Lebensdauerbemessung von Stahlbetonbauwerken – Zuverlässigkeitsbetrachtungen zur wirksamen Vermeidung von Bewehrungskorrosion*. PhD thesis, RWTH Aachen, 2000.

[7] W. Breit, C. Dauberschmidt, C. Gehlen, C. Sodeikat, A. Taffe, and U. Wiens, "Zum Ansatz eines kritischen Chloridgehaltes bei Stahlbetonbauwerken," *Beton- und Stahlbetonbau*, vol. 106, pp. 290–298, 5 2011.

[8] S. von Greve-Dierfeld and C. Gehlen, "Performance based durability design, carbonation part 1 – benchmarking of european present design rules," *Structural Concrete*, vol. 17, pp. 309–328, 9 2016.

[9] S. von Greve-Dierfeld and C. Gehlen, "Performance-based durability design, carbonation part 2 – classification of concrete," *Struct. Concr.*, vol. 17, pp. 523–532, 12 2016.

[10] S. von Greve-Dierfeld and C. Gehlen, "Performance-based durability design, carbonation, part 3: Psf approach and a proposal for the revision of deemed-to-satisfy rules," *Structural Concrete*, vol. 17, pp. 718–728, 12 2016.

[11] D. Sanio, J. Löschmann, P. Mark, and M. A. Ahrens, "Bauwerksmessungen versus Rechenkonzepte zur Beurteilung von Spannstahlermüdung in Betonbrücken," *Bautechnik*, vol. 95, pp. 99–110, 2 2018.

[12] S. Küttenbaum, S. Maack, T. Braml, A. Taffe, and T. Strübing, "Bewertung von Bestandsbauwerken mit gemessenen Daten, Teil 2," *Beton- und Stahlbetonbau*, vol. 116, pp. 183–199, 3 2021.

[13] "BASt - Brückenstatistik." `https://www.bast.de/DE/Statistik/Bruecken/Brueckenstatistik.html`, 2021. Accessed: 2023-01-06.

[14] C. Gehlen, S. V. Greve-Dierfeld, and K. Osterminski, "Modelling of ageing and corrosion processes in reinforced concrete structures," *Non-Destructive Evaluation of Reinforced Concrete Structures: Deterioration Processes and Standard Test Methods*, pp. 57–81, 1 2010.

[15] K. Aki, "Analysis of the seismic coda of local earthquakes as scattered waves," *Journal of Geophysical Research*, vol. 74, pp. 615–631, 1 1969.

[16] K. Aki and B. Chouet, "Origin of coda waves: Source, attenuation, and scattering effects," *Journal of Geophysical Research*, vol. 80, pp. 3322–3342, 8 1975.

[17] R. Snieder, A. Grêt, H. Douma, and J. Scales, "Coda wave interferometry for estimating nonlinear behavior in seismic velocity," *Science*, vol. 295, pp. 2253–2255, 3 2002.

[18] R. Snieder, "The theory of coda wave interferometry," *Pure and Applied Geophysics*, vol. 163, pp. 455–473, 3 2006.

[19] T. Planès and E. Larose, "A review of ultrasonic Coda Wave Interferometry in concrete," *Cement and Concrete Research*, vol. 53, pp. 248–255, 2013.

[20] C. Sens-Schönfelder and U. Wegler, "Passive image interferometry and seasonal variations of seismic velocities at merapi volcano, indonesia," *Geophysical Research Letters*, vol. 33, no. 21, p. 547, 2006.

[21] "DFG - GEPRIS - FOR 2825: Concrete Damage Assessment by Coda Waves (CoDA)." `https://gepris.dfg.de/gepris/projekt/398216472?language=en`, 2019. mboxAccessed: 2023-01-06.

[22] E. Niederleithinger, N. Epple, D. F. Barroso, F. Clauß, M. A. Ahrens, and P. Mark, "Ultraschallbasierte Überwachung von Stahl- und Spannbetonkonstruktionen – erste Ergebnisse der DFG-Forschungsgruppe 2825 CoDA," *4. Brückenkolloquium*, pp. 473–480, 2020.

[23] G. Vu, F. Diewald, J. J. Timothy, C. Gehlen, and G. Meschke, "Reduced order multiscale simulation of diffuse damage in concrete," *Materials*, vol. 14, p. 3830, 7 2021.

[24] V. Holla, G. Vu, J. J. Timothy, F. Diewald, C. Gehlen, and G. Meschke, "Computational generation of virtual concrete mesostructures," *Materials*, vol. 14, p. 3782, 7 2021.

[25] C. Finger, L. Saydak, G. Vu, J. J. Timothy, G. Meschke, and E. H. Saenger, "Sensitivity of ultrasonic coda wave interferometry to material damage — observations from a virtual concrete lab," *Materials*, vol. 14, p. 4033, 7 2021.

[26] E. H. Saenger, C. Finger, S. Karimpouli, and P. Tahmasebi, "Single-station coda wave interferometry: A feasibility study using machine learning," *Materials*, vol. 14, no. 13, 2021.

[27] G. Vu, J. J. Timothy, D. S. Singh, L. A. Saydak, E. H. Saenger, G. Meschke, and J. A. F. O. Correia, "Numerical simulation-based damage identification in concrete," *Modelling*, vol. 2, pp. 355–369, 8 2021.

[28] F. Clauß, N. Epple, M. A. Ahrens, E. Niederleithinger, and P. Mark, "Comparison of experimentally determined two-dimensional strain fields and mapped ultrasonic data processed by coda wave interferometry," *Sensors*, vol. 20, p. 4023, 7 2020.

[29] E. Niederleithinger, X. Wang, N. Epple, T. Schumacher, S. Ahmed, and P. Klikowicz, "Ultrasonic coda wave monitoring of concrete structures: First experiences with large-scale experiments and real structures," *Proceedings of the 10th International Conference on Bridge Maintenance, Safety and Management, IABMAS 2020*, pp. 2525–2531, 4 2021.

[30] F. Clauß, M. A. Ahrens, and P. Mark, "Thermo-mechanical experiments on reinforced concrete beams: Assessing thermal, mechanical, and mixed impacts on fiber optic measurements," *Structural Concrete*, vol. 23, pp. 3521–3537, 12 2022.

[31] F. Clauß, N. Epple, M. A. Ahrens, E. Niederleithinger, and P. Mark, "Correlation of load-bearing behavior of reinforced concrete members and velocity changes of coda waves," *Materials*, vol. 15, p. 738, 1 2022.

[32] Y. Zhang, T. Planès, E. Larose, A. Obermann, C. Rospars, and G. Moreau, "Diffuse ultrasound monitoring of stress and damage development on a

15-ton concrete beam," *The Journal of the Acoustical Society of America*, vol. 139, p. 1691, 4 2016.

[33] T. Planès, E. Larose, V. Rossetto, and L. Margerin, "Locadiff: Locating a weak change with diffuse ultrasound," *AIP Conference Proceedings*, vol. 1511, p. 405, 1 2013.

[34] S. Grabke, F. Clauß, K. U. Bletzinger, M. A. Ahrens, P. Mark, and R. Wüchner, "Damage detection at a reinforced concrete specimen with coda wave interferometry," *Materials*, vol. 14, p. 5013, 9 2021.

[35] S. Grabke, K. U. Bletzinger, and R. Wüchner, "Development of a finite element-based damage localization technique for concrete by applying coda wave interferometry," *Engineering Structures*, vol. 269, p. 114585, 10 2022.

[36] R. Springenschmid, *Betontechnologie für die Praxis*. Beuth Verlag, 2 ed., 2018.

[37] G. Wischers, "Einfluss der Zusammensetzung des Betons auf seine Frühfestigkeit," *Beton*, vol. 13, pp. 427–432, 1963.

[38] "DIN EN 197-1:2011-11, Zement - Teil 1: Zusammensetzung, Anforderungen und Konformitätskriterien von Normalzement; Deutsche Fassung EN 197-1:2011," Standard, Deutsches Institut für Normung, Berlin, 2011.

[39] "ASTM C150-07: Specification for Portland Cement," Standard, ASTM International, West Conshohocken, PA, USA, 12 2012.

[40] B. Pichler and C. Hellmich, "Upscaling quasi-brittle strength of cement paste and mortar: A multi-scale engineering mechanics model," *Cement and Concrete Research*, vol. 41, pp. 467–476, 5 2011.

[41] G. Constantinides and F. J. Ulm, "The effect of two types of c-s-h on the elasticity of cement-based materials: Results from nanoindentation and micromechanical modeling," *Cement and Concrete Research*, vol. 34, pp. 67–80, 1 2004.

[42] K. Velez, S. Maximilien, D. Damidot, G. Fantozzi, and F. Sorrentino, "Determination by nanoindentation of elastic modulus and hardness of pure constituents of portland cement clinker," *Cement and Concrete Research*, vol. 31, pp. 555–561, 4 2001.

[43] C. J. Haecker, E. J. Garboczi, J. W. Bullard, R. B. Bohn, Z. Sun, S. P. Shah, and T. Voigt, "Modeling the linear elastic properties of portland cement paste," *Cement and Concrete Research*, vol. 35, pp. 1948–1960, 10 2005.

[44] N. Roussel, H. Bessaies-Bey, S. Kawashima, D. Marchon, K. Vasilic, and R. Wolfs, "Recent advances on yield stress and elasticity of fresh cement-based materials," *Cement and Concrete Research*, vol. 124, p. 105798, 10 2019.

[45] J. Schön, *Physical properties of rocks : fundamentals and principles of petrophysics*, vol. 18. Elsevier B.V., 1996.

[46] I. Lillamand, J. F. Chaix, M. A. Ploix, and V. Garnier, "Acoustoelastic effect in concrete material under uniaxial compressive loading," *NDT & E International*, vol. 43, pp. 655–660, 11 2010.

[47] S. Popovics, J. L. Rose, and J. S. Popovics, "The behaviour of ultrasonic pulses in concrete," *Cement and Concrete Research*, vol. 20, pp. 259–270, 3 1990.

[48] B. C. Kim and J. Y. Kim, "Characterization of ultrasonic properties of concrete," *Mechanics Research Communications*, vol. 36, pp. 207–214, 3 2009.

[49] K. L. Scrivener and K. M. Nemati, "The percolation of pore space in the cement paste/aggregate interfacial zone of concrete," *Cement and Concrete Research*, vol. 26, pp. 35–40, 1 1996.

[50] K. L. Scrivener, A. K. C. Lyon, and F. P. Laugesen, "The interfacial transition zone (itz) between cement paste and aggregate in concrete," *Interface Science*, vol. 12, pp. 411–421, 2004.

[51] Z. Jia, Y. Han, Y. Zhang, C. Qiu, C. Hu, and Z. Li, "Quantitative characterization and elastic properties of interfacial transition zone around coarse aggregate in concrete," *Journal Wuhan University of Technology, Materials Science Edition*, vol. 32, pp. 838–844, 8 2017.

[52] T. Mori and K. Tanaka, "Average stress in matrix and average elastic energy of materials with misfitting inclusions," *Acta metallurgica*, vol. 21, no. 5, pp. 571–574, 1973.

[53] E. C. David and R. W. Zimmerman, "Pore structure model for elastic wave velocities in fluid-saturated sandstones," *J. Geophys. Res.*, vol. 117, 7 2012.

[54] C. Grosse, *Quantitative zerstörungsfreie Prüfung von Baustoffen mittels Schallemissionsanalyse und Ultraschall*. PhD thesis, Universität Stuttgart, 2 2013.

[55] H. Yildirim and O. Sengul, "Modulus of elasticity of substandard and normal concretes," *Construction and Building Materials*, vol. 25, pp. 1645–1652, 4 2011.

[56] K. Wesche and W. Manns, "Résultats d'une enquête internationale sur la détermination du module d'élasticité du béton en compression," *Materials and Structures*, vol. 3, no. 3, pp. 179–196, 1970.

[57] W. Voigt, *Lehrbuch der Kristallphysik (mit Ausschluss der Kristalloptik)*, vol. 1. Leipzig und Berlin: B.G Teubner, 1910.

[58] A. Reuss, "Berechnung der Fließgrenze von Mischkristallen auf Grund der Plastizitätsbedingung für Einkristalle," *ZAMM - Journal of Applied Mathematics and Mechanics / Zeitschrift für Angewandte Mathematik und Mechanik*, vol. 9, pp. 49–58, 1 1929.

[59] H. He and K. Senetakis, "A study of wave velocities and poisson ratio of recycled concrete aggregate," *Soils and Foundations*, vol. 56, pp. 593–607, 8 2016.

[60] K. Alexandrov and T. Ryzhova, "Elastic properties of rock-forming minerals: III Feldspars," *Bulletin of the Academy of Sciences, USSR, Geophysical Series*, vol. 10, pp. 129–131, 1962.

[61] H. Gebrande, "Rock forming minerals: Datasheet from Landolt-Börnstein - Group V Geophysics." `https://materials.springer.com/bp/docs/978-3-540-31365-6`. Accessed 2022-06-25.

[62] A. Machner, M. Zajac, M. B. Haha, K. O. Kjellsen, M. R. Geiker, and K. D. Weerdt, "Portland metakaolin cement containing dolomite or limestone – similarities and differences in phase assemblage and compressive strength," *Construction and Building Materials*, vol. 157, pp. 214–225, 12 2017.

[63] P. Shafigh, H. B. Mahmud, M. Z. Jumaat, and M. Zargar, "Agricultural wastes as aggregate in concrete mixtures – a review," *Construction and Building Materials*, vol. 53, pp. 110–117, 2 2014.

[64] M. Thiedeitz, B. Ostermaier, and T. Kränkel, "Rice husk ash as an additive in mortar – contribution to microstructural, strength and durability performance," *Resources, Conservation and Recycling*, vol. 184, p. 106389, 9 2022.

[65] "ASTM C494/C494M-17: Specification for Chemical Admixtures for Concrete," Standard, ASTM International, West Conshohocken, PA, USA, 3 2020.

[66] T. C. Powers, "A working hypothesis for further studies of frost resistance of concrete," *Journal Proceedings*, vol. 41, pp. 245–272, 1 1945.

[67] R. Auberg, *Zuverlässige Prüfung des Frost- und an-Widerstands von Beton mit dem CDF- und CIF-Test.* Shaker, als ms. gedr. ed., 1999.

[68] M. Mueller, *Frost-Tausalz-Angriff auf Beton - Neue Erkenntnisse zum Schadensmechanismus.* PhD thesis, Bauhaus Universität Weimar, 2021.

[69] D. W. Ho and R. K. Lewis, "Carbonation of concrete and its prediction," *Cement and Concrete Research*, vol. 17, pp. 489–504, 5 1987.

[70] A. V. Saetta, B. A. Schrefler, and R. V. Vitaliani, "The carbonation of concrete and the mechanism of moisture, heat and carbon dioxide flow through porous materials," *Cement and Concrete Research*, vol. 23, pp. 761–772, 7 1993.

[71] L. L. Fritz Hunkeler, "Anforderungen an den Karbonatisierungswiderstand von Betonen," *Eidgenössisches Departement für Umwelt, Verkehr, Energie und Kommunikation, Bundesamt für Strassen*, 2012.

[72] W. Breit, *Critical corrosion-inducing chloride content - state of the art and new investigation results.* Düsseldorf: Verlag Bau + Technik, 2001.

[73] U. Angst, B. Elsener, C. K. Larsen, and Øystein Vennesland, "Critical chloride content in reinforced concrete — a review," *Cement and Concrete Research*, vol. 39, pp. 1122–1138, 12 2009.

[74] H. Grube, *Ursachen des Schwindens von Beton und Auswirkungen auf Betonbauteile*, vol. 52 of *Schriftenreihe der Zementindustrie.* Düsseldorf: Beton-Verl., 1991.

[75] L. Meyer, *Zum Einfluss der Kontaktzone bei der Modellierung des Elastizitätsmoduls von Beton.* PhD thesis, RWTH Aachen, 2007.

[76] J. Krautkrämer and H. Krautkrämer, “Werkstoffprüfung mit Ultraschall,” *Springer Verlag, Berlin Heidelberg*, 1986.

[77] Ö. Yilmaz, *Seismic Data Analysis*. Society of Exploration Geophysicists, 1 2001.

[78] “PyCMG - Concrete Mesostructure Generator.” `https://github.com/jtimo/pycmg.git`, 2021. Accessed: 2023-01-13.

[79] P. Anugonda, J. S. Wiehn, and J. A. Turner, “Diffusion of ultrasound in concrete,” *Ultrasonics*, vol. 39, pp. 429–435, 10 2001.

[80] T. Planès, E. Larose, L. Margerin, V. Rossetto, and C. Sens-Schönfelder, “Decorrelation and phase-shift of coda waves induced by local changes: multiple scattering approach and numerical validation,” *Waves in Random and Complex Media*, vol. 24, pp. 99–125, 4 2014.

[81] J. S. Popovics and J. L. Rose, “A survey of developments in ultrasonic nde of concrete,” *IEEE Transactions on Ultrasonics, Ferroelectrics, and Frequency Control*, vol. 41, pp. 140–143, 1994.

[82] G. Poupinet, W. L. Ellsworth, and J. Frechet, “Monitoring velocity variations in the crust using earthquake doublets: An application to the calaveras fault, california,” *Journal of Geophysical Research: Solid Earth*, vol. 89, no. B7, pp. 5719–5731, 1984.

[83] R. Snieder, “Imaging and averaging in complex media,” *Diffuse Waves in Complex Media*, pp. 405–454, 1999.

[84] G. Poupinet, W. L. Ellsworth, and J. Frechet, “Monitoring velocity variations in the crust using earthquake doublets: An application to the calaveras fault, california,” *Journal of Geophysical Research: Solid Earth*, vol. 89, pp. 5719–5731, 7 1984.

[85] P. M. Roberts, M. C. Fehler, P. A. Johnson, and W. S. Phillips, “Active doublet method for measuring small changes in physical properties,” Patent, Los Alamos National Lab. (LANL), USA, 1 1994.

[86] C. Sens-Schönfelder and U. Wegler, “Passive image interferemetry and seasonal variations of seismic velocities at Merapi Volcano, Indonesia,” *Geophysical Research Letters*, vol. 33, nov 2006.

[87] E. Niederleithinger and C. Wunderlich, “Influence of small temperature variations on the ultrasonic velocity in concrete,” *AIP Conference Proceedings*, vol. 1511, pp. 390–397, 2013.

[88] P. Fröjd and P. Ulriksen, “Detecting damage events in concrete using diffuse ultrasound structural health monitoring during strong environmental variations,” *Structural Health Monitoring*, vol. 17, no. 2, pp. 410–419, 2018.

[89] D. P. Schurr, J. Y. Kim, K. G. Sabra, and L. J. Jacobs, “Damage detection in concrete using coda wave interferometry,” *NDT & E International*, vol. 44, pp. 728–735, 12 2011.

[90] A. Quiviger, C. Payan, J. F. Chaix, V. Garnier, and J. Salin, “Effect of the presence and size of a real macro-crack on diffuse ultrasound in concrete,” *NDT and E International*, vol. 45, pp. 128–132, jan 2012.

[91] E. Niederleithinger, J. Wolf, F. Mielentz, H. Wiggenhauser, and S. Pirskawetz, “Embedded ultrasonic transducers for active and passive concrete monitoring,” *Sensors*, vol. 15, no. 5, pp. 9756–9772, 2015.

[92] H. Jiang, J. Zhang, and R. Jiang, “Stress Evaluation for Rocks and Structural Concrete Members through Ultrasonic Wave Analysis: Review,” *Journal of Materials in Civil Engineering*, vol. 29, no. 10, p. 04017172, 2017.

[93] J. Chakraborty, A. Katunin, P. Klikowicz, and M. Salamak, “Early Crack Detection of Reinforced Concrete Structure Using Embedded Sensors,” *Sensors*, vol. 19, p. 3879, sep 2019.

[94] A. Narayanan, Ali Sheikh Ali, B. Delsaute, C. Pierre, and A. Deraemaeker, “Automated and continuous monitoring of freeze-thaw damage in concrete using embedded piezoelectric transducers,” *Lecture Notes in Civil Engineering*, vol. 254 LNCE, pp. 843–851, 2023.

[95] S. Liu, Z. B. Bundur, J. Zhu, and R. D. Ferron, “Evaluation of self-healing of internal cracks in biomimetic mortar using coda wave interferometry,” *Cement and Concrete Research*, vol. 83, pp. 70–78, 5 2016.

[96] S. Mehdinia, K. S. T. Chopperla, A. Hafiz, T. Schumacher, and J. H. Ideker, “Ultrasonic coda wave monitoring of alkali-silica reactivity in concrete laboratory prisms,” *Materials Evaluation*, vol. 80, pp. 40–51, 10 2022.

[97] E. Ohdaira and N. Masuzawa, “Water content and its effect on ultrasound propagation in concrete - the possibility of NDE,” *Ultrasonics*, vol. 38, no. 1, pp. 546–552, 2000.

[98] S. Popovics, "Effects of uneven moisture distribution on the strength of and wave velocity in concrete," *Ultrasonics*, vol. 43, pp. 429–434, may 2005.

[99] F. D. Murnaghan, "Finite deformations of an elastic solid," *American Journal of Mathematics*, vol. 59, p. 235, 4 1937.

[100] D. S. Hughes and J. L. Kelly, "Second-order elastic deformation of solids," *Physical Review*, vol. 92, p. 1145, 12 1953.

[101] Y. Zhang, O. Abraham, F. Grondin, A. Loukili, V. Tournat, A. L. Duff, B. Lascoup, and O. Durand, "Study of stress-induced velocity variation in concrete under direct tensile force and monitoring of the damage level by using thermally-compensated coda wave interferometry," *Ultrasonics*, vol. 52, pp. 1038–1045, 12 2012.

[102] Y. Zhang, O. Abraham, V. Tournat, A. L. Duff, B. Lascoup, A. Loukili, F. Grondin, and O. Durand, "Validation of a thermal bias control technique for Coda Wave Interferometry (CWI)," *Ultrasonics*, vol. 53, pp. 658–664, 3 2013.

[103] F. Diewald, N. Epple, T. Kraenkel, C. Gehlen, and E. Niederleithinger, "Impact of external mechanical loads on coda waves in concrete," *Materials*, vol. 15, p. 5482, 8 2022.

[104] A. Hafiz and T. Schumacher, "Monitoring of stresses in concrete using ultrasonic coda wave comparison technique," *Journal of Nondestructive Evaluation*, vol. 37, p. 73, 2018.

[105] B. Hilloulin, Y. Zhang, O. Abraham, A. Loukili, F. Grondin, O. Durand, and V. Tournat, "Small crack detection in cementitious materials using nonlinear coda wave modulation," *NDT & E International*, vol. 68, pp. 98–104, 12 2014.

[106] A. Deraemaeker and C. Dumoulin, "Embedding ultrasonic transducers in concrete: A lifelong monitoring technology," *Construction and Building Materials*, vol. 194, pp. 42–50, 1 2019.

[107] E. Larose, J. D. Rosny, L. Margerin, D. Anache, P. Gouedard, M. Campillo, and B. V. Tiggelen, "Observation of multiple scattering of khz vibrations in a concrete structure and application to monitoring weak changes," *Physical Review E - Statistical, Nonlinear, and Soft Matter Physics*, vol. 73, p. 016609, 1 2006.

[108] S. Mehdinia, C. K. S. Teja, A. Hafiz, T. Schumacher, and J. H. Ideker, "Ultrasonic coda wave monitoring of alkali-silica reactivity in concrete laboratory prisms," *Materials Evaluation*, vol. 80, 2022.

[109] S. C. Stähler, C. Sens-Schönfelder, and E. Niederleithinger, "Monitoring stress changes in a concrete bridge with coda wave interferometry," *The Journal of the Acoustical Society of America*, vol. 129, p. 1945, 4 2011.

[110] J.-B. Legland, Y. Zhang, O. Abraham, O. Durand, and V. Tournat, "Evaluation of crack status in a meter-size concrete structure using the ultrasonic nonlinear coda wave interferometry," *The Journal of the Acoustical Society of America*, vol. 142, p. 2233, 10 2017.

[111] Q.-Y. Wang, F. Brenguier, M. Campillo, A. Lecointre, T. Takeda, and Y. Aoki, "Seasonal crustal seismic velocity changes throughout Japan," *Journal of Geophysical Research: Solid Earth*, vol. 122, no. 10, pp. 7987–8002, 2017.

[112] K.-F. Feng, H.-H. Huang, Y.-J. Hsu, and Y.-M. Wu, "Controls on seasonal variations of crustal seismic velocity in taiwan using single-station cross-component analysis of ambient noise interferometry," *Journal of Geophysical Research: Solid Earth*, vol. 126, no. 11, p. e2021JB022650, 2021.

[113] L. Ermert, E. Cabral-Cano, E. Chaussard, D. Solano-Rojas, L. Quintanar, D. Morales Padilla, E. A. Fernandez-Torres, and M. A. Denolle, "Probing environmental and tectonic changes underneath ciudad de méxico with the urban seismic field," *EGUsphere*, pp. 1–33, 2023.

[114] T. Clements and M. Denolle, "The seismic signature of California's earthquakes, droughts, and floods," *Journal of Geophysical Research: Solid Earth*, vol. XX, no. X, p. XX, 2023.

[115] H. Sun and J. Zhu, "Thermal modulation of nonlinear ultrasonic wave for concrete damage evaluation," *The Journal of the Acoustical Society of America*, vol. 145, p. EL405, 5 2019.

[116] C. Sens-Schönfelder and U. Wegler, "Passive image interferemetry and seasonal variations of seismic velocities at Merapi volcano, Indonesia," *Geophysical Research Letters*, vol. 33, 11 2006.

[117] C. Hadziioannou, E. Larose, and O. Coutant, "Stability of monitoring weak changes in multiply scattering media with ambient noise correlation: Laboratory experiments," *The Journal of the Acoustical Society of America*, vol. 125, p. 3688, 2009.

[118] C. Yuan, J. Bryan, and M. Denolle, "Numerical comparison of time-, frequency-and wavelet-domain methods for coda wave interferometry," *Geophysical Journal International*, vol. 226, no. 2, pp. 828–846, 2021.

[119] E. Niederleithinger, X. Wang, M. Herbrand, and M. Müller, "Processing ultrasonic data by coda wave interferometry to monitor load tests of concrete beams," *Sensors*, vol. 18, no. 6, 2018.

[120] S. Mak, L. S. Chan, A. M. Chandler, and R. C. Koo, "Coda Q estimates in the Hong Kong region," *Journal of Asian Earth Sciences*, vol. 24, pp. 127–136, 10 2004.

[121] M. Calvet and L. Margerin, "Lapse-time dependence of coda Q: Anisotropic multiple-scattering models and application to the Pyrenees," *Bulletin of the Seismological Society of America*, vol. 103, no. 3, pp. 1993–2010, 2013.

[122] Y. Lee, S. T. Yi, M. S. Kim, and J. K. Kim, "Evaluation of a basic creep model with respect to autogenous shrinkage," *Cement and Concrete Research*, vol. 36, pp. 1268–1278, 7 2006.

[123] L. Wu, N. Farzadnia, C. Shi, Z. Zhang, and H. Wang, "Autogenous shrinkage of high performance concrete: A review," *Construction and Building Materials*, vol. 149, pp. 62–75, 9 2017.

[124] J. K. Kim and C. S. Lee, "Prediction of differential drying shrinkage in concrete," *Cement and Concrete Research*, vol. 28, pp. 985–994, 7 1998.

[125] A. Fabbri, J. Corvisier, A. Schubnel, F. Brunet, B. Goffé, G. Rimmele, and V. Barlet-Gouédard, "Effect of carbonation on the hydro-mechanical properties of portland cements," *Cement and Concrete Research*, vol. 39, pp. 1156–1163, 12 2009.

[126] P. Fröjd and P. Ulriksen, "Frequency selection for coda wave interferometry in concrete structures," *Ultrasonics*, vol. 80, pp. 1–8, 9 2017.

[127] D. F. Barroso, N. Epple, E. Niederleithinger, F. M. Agugliaro, and E. Salmerón-Manzano, "A portable low-cost ultrasound measurement device for concrete monitoring," *Inventions*, vol. 6, p. 36, 5 2021.

[128] H. Sogbossi, J. Verdier, and S. Multon, "Permeability and damage of partially saturated concrete exposed to elevated temperature," *Cement and Concrete Composites*, vol. 109, p. 103563, 5 2020.

[129] I. Hager, "Behaviour of cement concrete at high temperature," *Bulletin of the Polish Academy of Sciences: Technical Sciences*, vol. 61, pp. 145–154, 3 2013.

[130] F. Grondin, H. Dumontet, A. B. Hamida, and H. Boussa, "Micromechanical contributions to the behaviour of cement-based materials: Two-scale modelling of cement paste and concrete in tension at high temperatures," *Cement and Concrete Composites*, vol. 33, pp. 424–435, 3 2011.

[131] N. Hearn, "Effect of shrinkage and load-induced cracking on water permeability of concrete," *ACI Materials Journal*, vol. 96, pp. 234–241, 3 1999.

[132] J. C. Mindeguia, P. Pimienta, A. Noumowé, and M. Kanema, "Temperature, pore pressure and mass variation of concrete subjected to high temperature - experimental and numerical discussion on spalling risk," *Cement and Concrete Research*, vol. 40, pp. 477–487, 3 2010.

[133] Z. P. Bažant, "Constitutive equation for concrete creep and shrinkage based on thermodynamics of multiphase systems," *Matériaux et Constructions*, vol. 3, pp. 3–36, 1 1970.

[134] Q. Zhou and F. P. Glasser, "Thermal stability and decomposition mechanisms of ettringite at <120 °C," *Cement and Concrete Research*, vol. 31, pp. 1333–1339, 9 2001.

[135] "DIN EN 1992-1-1:2011-01, Eurocode 2: Bemessung und Konstruktion von Stahlbeton- und Spannbetontragwerken - Teil 1-1: Allgemeine Bemessungsregeln und Regeln für den Hochbau," Standard, Deutsches Institut für Normung, Berlin, 2011.

[136] L. Yang, H. Jia, L. Han, H. Zhang, and L. Tang, "Hysteresis in the ultrasonic parameters of saturated sandstone during freezing and thawing and correlations with unfrozen water content," *Journal of Rock Mechanics and Geotechnical Engineering*, vol. 13, no. 5, pp. 1078–1092, 2021.

[137] C. Wunderlich and E. Niederleithinger, "Evaluation of temperature influence on ultrasound velocity in concrete by coda wave interferometry," *RILEM Bookseries*, vol. 6, pp. 227–232, 2012.

[138] R. L. Weaver, "Diffuse elastic waves at a free surface," *The Journal of the Acoustical Society of America*, vol. 78, no. 1, pp. 131–136, 1985.

[139] C. Donaldson, T. Winder, C. Caudron, and R. S. White, "Crustal seismic velocity responds to a magmatic intrusion and seasonal loading in iceland's northern volcanic zone," *Science advances*, vol. 5, no. 11, p. 6642, 2019.

[140] P. J. Monteiro, S. A. Miller, and A. Horvath, "Towards sustainable concrete," *Nature Materials*, vol. 16, pp. 698–699, 6 2017.

[141] H. Zhan, H. Jiang, C. Zhuang, J. Zhang, and R. Jiang, "Estimation of stresses in concrete by using coda wave interferometry to establish an acoustoelastic modulus database," *Sensors*, vol. 20, p. 4031, 7 2020.

[142] B. Zhong and J. Zhu, "Applications of stretching technique and time window effects on ultrasonic velocity monitoring in concrete," *Applied Sciences*, vol. 12, p. 7130, 7 2022.

[143] H. Jiang, M. Asce, H. Zhan, J. Zhang, R. Jiang, C. Zhuang, and P. Fan, "Detecting stress changes and damage in full-size concrete t-beam and slab with ultrasonic coda waves," *Journal of Structural Engineering*, vol. 147, p. 04021140, 7 2021.

[144] S. Popovics and J. S. Popovics, "Effect of stresses on the ultrasonic pulse velocity in concrete," *Materials and Structures*, vol. 24, pp. 15–23, 1 1991.

[145] C. Payan, V. Garnier, J. Moysan, and P. A. Johnson, "Determination of third order elastic constants in a complex solid applying coda wave interferometry," *Applied Physics Letters*, vol. 94, no. 1, pp. 1–3, 2009.

[146] J. G. Van Mier, *Fracture processes of concrete*, vol. 12. CRC Press, 1996.

[147] K. M. Nemati, P. J. Monteiro, and K. L. Scrivener, "Analysis of compressive stress-induced cracks in concrete," *ACI Materials Journal*, vol. 95, pp. 617–630, 1998.

[148] K. Wesche, "Baustoffe für uteile - Band 2: Beton, Mauerwerk (Nichtmetallisch-anorganische Stoffe): Herstellung, Eigenschaften, Verwendung, Dauerhaftigkeit," *Vieweg+Teubner Verlag*, vol. 2, 1993.

[149] M. H. Witt, *Untersuchungen zur Kerbtiefe bei Querscheinfugen in Betonfahrbahnen.* PhD thesis, Technical University of Munich, 2018.

[150] H. S. Müller and F. Acosta, *Time dependent effects of structural concrete: Basics for constitutive modelling towards the next generation of EN EC 2*, pp. 395–413. Ernst und Sohn, 2014.

[151] H. S. Müller and V. Kvitsel, "Kriechen und Schwinden von Beton," *Beton- und Stahlbetonbau*, vol. 97, pp. 8–19, 1 2002.

[152] F. Diewald, L. Irbe, T. Kraenkel, A. Machner, and C. Gehlen, "Monitoring early cement hydration with coda wave interferometry," *Proceedings of the 13th International Workshop on Structural Health Monitoring*, p. 5482, 2022.

[153] P. J. M. Monteiro, S. A. Miller, and A. Horvath, "Towards sustainable concrete," *Nature materials*, vol. 16, no. 7, pp. 698–699, 2017.

[154] K. Aki, "Theory of earthquake prediction with special reference to monitoring of the quality factor of lithosphere by the coda method," in *Pract Appr to Earthquake Pred and Warning* (C. Kisslinger and T. Rikitake, eds.), vol. 72, pp. 219–230, Springer Netherlands, 1985.

[155] O. I. Lobkis and R. L. Weaver, "Coda-wave interferometry in finite solids: recovery of p-to-s conversion rates in an elastodynamic billiard," *Phys rev letters*, vol. 90, no. 25 Pt 1, p. 254302, 2003.

[156] D. Jansen, F. Goetz-Neunhoeffer, B. Lothenbach, and J. Neubauer, "The early hydration of ordinary portland cement (opc): An approach comparing measured heat flow with calculated heat flow from qxrd," *Cement and Concrete Research*, vol. 42, no. 1, pp. 134–138, 2012.

[157] H. Taylor, *Cement chemistry*. Thomas Telford Publishing, 2 ed., 1997.

[158] C. Hesse, F. Goetz-Neunhoeffer, and J. Neubauer, "A new approach in quantitative in-situ xrd of cement pastes: Correlation of heat flow curves with early hydration reactions," *Cement and Concrete Research*, vol. 41, pp. 123–128, 01 2011.

[159] B. Lothenbach and M. Zajac, "Application of thermodynamic modelling to hydrated cements," *Cement and Concrete Research*, vol. 123, p. 105779, 2019.

[160] B. Lothenbach, D. A. Kulik, T. Matschei, M. Balonis, L. Baquerizo, B. Dilnesa, G. D. Miron, and R. J. Myers, "Cemdata18: A chemical thermodynamic database for hydrated portland cements and alkali-activated materials," *Cement and Concrete Research*, vol. 115, pp. 472–506, 2019.

[161] L. J. Parrot and D. C. Killoh, "Prediction of cement hydration," *Br. Ceram. Proc.*, vol. 35, pp. 41–53, 1984.

[162] D. A. Kulik, T. Wagner, S. V. Dmytrieva, G. Kosakowski, F. F. Hingerl, K. V. Chudnenko, and U. R. Berner, "Gem-selektor geochemical modeling package: Revised algorithm and gems3k numerical kernel for coupled simulation codes," *Comp Geosc*, vol. 17, no. 1, pp. 1–24, 2013.

[163] D. A. Kulik, F. Winnefeld, A. Kulik, G. D. Miron, and B. Lothenbach, "Cemgems – an easy-to-use web application for thermodynamic modeling of cementitious materials," *RILEM Technical Letters*, vol. 6, pp. 36–52, 2021.

[164] B. Lothenbach and F. Winnefeld, "Thermodynamic modelling of the hydration of portland cement," *Cement and Concrete Research*, vol. 36, no. 2, pp. 209–226, 2006.

[165] W. Breit, "Bauwerküberwachung mit Hilfe von nachträglich installierten Feuchtesensoren (Multi-Ring-Elektroden)," Report, Institut für Bauforschung Aachen, 2012.

[166] N. Epple, D. F. Barroso, E. Niederleithinger, I. Hindersmann, C. Sodeikat, and R. Groschup, "From the lab to the structure: Monitoring of a german road bridge using embedded ultrasonic transducers and coda waves," *Lecture Notes in Civil Engineering*, vol. 254 LNCE, pp. 824–832, 2023.

[167] "Future of monitoring civil engineering structures." `https://chat.openai.com/chat/309bbe29-19ad-4a8e-a1bb-bb4845d233e6`, 2023. Accessed: 2023-01-13.

Verzeichnis der in der Schriftenreihe des Deutschen Ausschusses für Stahlbeton – DAfStb – seit 1945 erschienenen Hefte

Heft

100: Versuche an Stahlbetonbalken zur Bestimmung der Bewehrungsgrenze.
Von *W. Gehler, H. Amos* und *E. Friedrich.*
Die Ergebnisse der Versuche und das Dresdener Rechenverfahren für den plastischen Betonbereich (1949).
Von *W. Gehler.* 9,70 EUR

101: Versuche zur Ermittlung der Rissbildung und der Widerstandsfähigkeit von Stahlbetonplatten mit verschiedenen Bewehrungsstählen bei stufenweise gesteigerter Last.
Von *O. Graf* und *K. Walz.*
Versuche über die Schwellzugfestigkeit von verdrillten Bewehrungsstählen.
Von *O. Graf* und *G. Weil.*
Versuche über das Verhalten von kalt verformten Baustählen beim Zurückbiegen nach verschiedener Behandlung der Proben.
Von *O. Graf* und *G. Weil.*
Versuche zur Ermittlung des Zusammenwirkens von Fertigbauteilen aus Stahlbeton für Decken (1948).
Von *H. Amos* und *W. Bochmann.* vergriffen

102: Beton und Zement im Seewasser (1950).
Von *A. Eckhardt* und *W. Kronsbein.* vergriffen

103: Die *n*-freien Berechnungsweisen des einfach bewehrten, rechteckigen Stahlbetonbalkens (1951).
Von *K. B. Haberstock.* vergriffen

104: Bindemittel für Massenbeton, Untersuchungen über hydraulische Bindemittel aus Zement, Kalk und Trass (1951).
Von *K. Walz.* vergriffen

105: Die Versuchsberichte des Deutschen Ausschusses für Stahlbeton (1951).
Von *O. Graf.* vergriffen

106: Berechnungstafeln für rechtwinklige Fahrbahnplatten von Straßenbrücken (1952). 7. neubearbeitete Auflage (1981).
Von *H. Rüsch.* vergriffen

107: Die Kugelschlagprüfung von Beton.
Von *K. Gaede.* vergriffen

108: Verdichten von Leichtbeton durch Rütteln (1952).
Von *K. Walz.* vergriffen

109: SO_3-Gehalt der Zuschlagstoffe (1952).
Von *K. Gaede.* 3,30 EUR

110: Ziegelsplittbeton (1952).
Von *K. Charisius, W. Drechsel* und *A. Hummel.* vergriffen

111: Modellversuche über den Einfluss der Torsionssteifigkeit bei einer Plattenbalkenbrücke (1952).
Von *G. Marten.* vergriffen

112: Eisenbahnbrücken aus Spannbeton (1953). 2. erweiterte Auflage (1961).
Von *R. Bührer.* 7,80 EUR

113: Knickversuche mit Stahlbetonsäulen.
Von *W. Gehler* und *A. Hütter.*
Festigkeit und Elastizität von Beton mit hoher Festigkeit (1954).
Von *O. Graf.* 9,10 EUR

Heft

114: Schüttbeton aus verschiedenen Zuschlagstoffen.
Von *A. Hummel* und *K. Wesche.*
Die Ermittlung der Kornfestigkeit von Ziegelsplitt und anderen Leichtbeton-Zuschlagstoffen (1954).
Von *A. Hummel.* vergriffen

115: Die Versuche der Bundesbahn an Spannbetonträgern in Kornwestheim (1954).
Von *U. Giehrach* und *C. Sättele.* 5,40 EUR

116: Verdichten von Beton mit Innenrüttlern und Rütteltischen, Güteprüfung von Deckensteinen (1954).
Von *K. Walz.* vergriffen

117: Gas- und Schaumbeton: Tragfähigkeit von Wänden und Schwinden.
Von *O. Graf* und *H. Schäffler.*
Kugelschlagprüfung von Porenbeton (1954).
Von *K. Gaede.* vergriffen

118: Schwefelverbindung in Schlackenbeton (1954).
Von *A. Stois, F. Rost, H. Zinnert* und *F. Henkel.* 6,90 EUR

119: Versuche über den Verbund zwischen Stahlbeton-Fertigbalken und Ortbeton.
Von *O. Graf* und *G. Weil.*
Versuche mit Stahlleichtträgern für Massivdecken (1955).
Von *G. Weil.* vergriffen

120: Versuche zur Festigkeit der Biegedruckzone (1955).
Von *H. Rüsch.* vergriffen

121: Gas- und Schaumbeton:
Versuche zur Schubsicherung bei Balken aus bewehrtem Gas- und Schaumbeton.
Von *H. Rüsch.*
Ausgleichsfeuchtigkeit von dampfgehärtetem Gas- und Schaumbeton.
Von *H. Schäffler.*
Versuche zur Prüfung der Größe des Schwindens und Quellens von Gas und Schaumbeton (1956).
Von *O. Graf* und *H. Schäffle.* vergriffen

122: Gestaltfestigkeit von Betonkörpern.
Von *K. Walz.*
Warmzerreißversuche mit Spannstählen.
Von *J. Dannenberg, H. Deutschmann* und *Melchior.*
Konzentrierte Lasteintragung in Beton (1957).
Von *W. Pohle.* 7,60 EUR

123: Luftporenbildende Betonzusatzmittel (1956).
Von *K. Walz.* vergriffen

124: Beton im Seewasser (Ergänzung zu Heft 102) (1956).
Von *A. Hummel* und *K. Wesche.* 2,70 EUR

125: Untersuchungen über Federgelenke (1957).
Von *K. Kammüller* und *O. Jeske.* vergriffen

126: SO_3-Gehalt der Zuschlagstoffe – Langzeitversuche (Ergänzung zu Heft 109). Eindringtiefe von Beton in Holzwolle-Leichtbauplatten (1957).
Von *K. Gaede.* 5,40 EUR

Heft

127: Witterungsbeständigkeit von Beton (1957)
Von *K. Walz.* 4,80 EUR

128: Kugelschlagprüfung von Beton (Einfluss des Betonalters) (1957).
Von *K. Gaede.* vergriffen

129: Stahlbetonsäulen unter Kurz- und Langzeitbelastung (1958).
Von *K. Gaede.* 12,90 EUR

130: Bruchsicherheit bei Vorspannung ohne Verbund (1959).
Von *H. Rüsch, K. Kordina* und *C. Zelger.* 5,40 EUR

131: Das Kriechen unbewehrten Betons (1958).
Von *O. Wagner.* vergriffen

132: Brandversuche mit starkbewehrten Stahlbetonsäulen.
Von *H. Seekamp.*
Widerstandsfähigkeit von Stahlbetonbauteilen und Stahlsteindecken bei Bränden (1959).
Von *M. Hannemann* und *H. Thoms.* vergriffen

133: Gas- und Schaumbeton:
Druckfestigkeit von dampfgehärtetem Gasbeton nach verschiedener Lagerung.
Von *H. Schäffler.*
Über die Tragfähigkeit von bewehrten Platten aus dampfgehärtetem Gas- und Schaumbeton.
Von *H. Schäffler.*
Untersuchung des Zusammenwirkens von Porenbeton mit Schwerbeton bei bewehrten Schwerbetonbalken mit seitlich angeordneten Porenbetonschalen (1959).
Von *H. Rüsch* und *E. Lassas.* 4,80 EUR

134: Über das Verhalten von Beton in chemisch angreifenden Wässern (1959).
Von *K. Seidel.* vergriffen

135: Versuche über die beim Betonieren an den Schalungen entstehenden Belastungen.
Von *O. Graf* und *K. Kaufmann.*
Druckfestigkeit von Beton in der oberen Zone nach dem Verdichten durch Innenrüttler.
Von *K. Walz* und *H. Schäffler.*
Versuche über die Verdichtung von Beton auf einem Rütteltisch in lose aufgesetzter und in aufgespannter Form (1960).
Von *J. Strey.* vergriffen

136: Gas- und Schaumbeton:
Versuche über die Verankerung der Bewehrung in Gasbeton.
Über das Kriechen von bewehrten Platten aus dampfgehärtetem Gas- und Schaumbeton (1960).
Von *H. Schäffler.* 11,20 EUR

137: Schubversuche an Spannbetonbalken ohne Schubbewehrung.
Von *H. Rüsch* und *G. Vigerust.*
Die Schubfestigkeit von Spannbetonbalken ohne Schubbewehrung (1960).
Von *G. Vigerust.* vergriffen

138: Über die Grundlagen des Verbundes zwischen Stahl und Beton (1961).
Von *G. Rehm.* vergriffen

139: Theoretische Auswertung von Heft 120 – Festigkeit der Biegedruckzone (1961).
Von *G. Scholz.* 5,80 EUR

Heft

252: Beständigkeit verschiedener Betonarten in Meerwasser und in sulfathaltigem Wasser (1975).
Von *H. T Schröder, O. Hallauer* und *W. Scholz.* 15,50 EUR

253: Spannbeton-Reaktordruckbehälter-Instrumentierung.
Von *J. Német* und *R. Angeli.*
Versuch zur Weiterentwicklung eines Setzdehnungsmessers (1975).
Von *C. Zelger.* 10,20 EUR

254: Festigkeit und Verformungsverhalten von Beton unter hohen zweiachsigen Dauerbelastungen und Dauerschwellbelastungen. Festigkeit und Verformungsverhalten von Leichtbeton, Gasbeton, Zementstein und Gips unter zweiachsiger Kurzzeitbeanspruchung (1976).
Von *D. Linse* und *A. Stegbauer.* 13,10 EUR

255: Zur Frage der zulässigen Rissbreite und der erforderlichen Betondeckung im Stahlbetonbau unter besonderer Berücksichtigung der Karbonatisierungstiefe des Betons (1976).
Von *P. Schiessl.* vergriffen

256: Wärme- und Feuchtigkeitsleitung in Beton unter Einwirkung eines Temperaturgefälles (1975).
Von *J. Hundt.* 15,80 EUR

257: Bruchsicherheitsberechnung von Spannbeton-Druckbehältern (1976).
Von *K. Schimmelpfennig.* 13,30 EUR

258: Hygrische Transportphänomene in Baustoffen (1976).
Von *K. Gertis, K. Kiesl, H. Werner* und *V. Wolfseher.* 13,10 EUR

259: Entwicklung eines integrierten Spannbetondruckbehälters für wassergekühlte Reaktoren (SBB Typ „Stern" mit Stützkessel) (1976).
Von *G. Jüptner, H. Kumpf, G. Molz, B. Neunert* und *O. Seidl.* 11,50 EUR

260: Studie zum Trag- und Verformungsverhalten von Stahlbeton (1976).
Von *J. Eibl* und *G. Ivànyi.* 26,80 EUR

261: Der Einfluss radioaktiver Strahlung auf die mechanischen Eigenschaften von Beton (1976).
Von *H. Hilsdorf, J. Kropp* und *H.-J. Koch.* 8,40 EUR

262: Experimentelle Bestimmung des räumlichen Spannungszustandes eines Reaktordruckbehältermodells (1976).
Von *R. Stöver.* 13,10 EUR

263: Bruchfestigkeit und Bruchverformung von Beton unter mehraxialer Belastung bei Raumtemperatur (1976).
Von *F. Bremer* und *F. Steinsdörfer.* 7,60 EUR

264 Spannbeton-Reaktordruckbehälter mit heißer Dichthaut für Druckwasserreaktoren (1976).
Von *A. Jungmann, H. Kopp, M. Gangl, J. Német, A. Nesitka, W. Walluschek-Wallfeld* und *J. Mutzl.* 10,70 EUR

265: Traglast von Stahlbetondruckgliedern unter schiefer Biegung (1976).
Von *K. Kordina, K. Rafla* und *O. Hjorth.* 11,80 EUR

266: Das Trag- und Verformungsverhalten von Stahlbetonbrückenpfeilern mit Rollenlagern (1976).
Von *K. Liermann.* 12,90 EUR

Heft

267: Zur Mindestbewehrung für Zwang von Außenwänden aus Stahlleichtbeton.
Von *F. S. Rostásy, R. Koch* und *F. Leonhardt.*
Versuche zum Tragverhalten von Druckübergreifungsstößen in Stahlbetonwänden (1976).
Von *F. Leonhardt, F. S. Rostásy* und *M. Patzak.* 15,00 EUR

268: Einfluss der Belastungsdauer auf das Verbundverhalten von Stahl in Beton (Verbundkriechen) (1976).
Von *L. Franke.* 8,60 EUR

269: Zugspannung und Dehnung in unbewehrten Betonquerschnitten bei exzentrischer Belastung (1976).
Von *H. G. Heilmann.* 15,50 EUR

270: Eine Formulierung des zweiaxialen Verformungs- und Bruchverhaltens von Beton und deren Anwendung auf die wirklichkeitsnahe Berechnung von Stahlbetonplatten (1976).
Von *J. Link.* 14,40 EUR

271: Untersuchungen an 20 Jahre alten Spannbetonträgern (1976).
Von *R. Bührer, K.-F. Müller, H. Martin* und *J. Ruhnau.* 13,10 EUR

272: Die Dynamische Relaxation und ihre Anwendung auf Spannbeton-Reaktordruckbehälter (1976).
Von *W. Zerna.* 13,70 EUR

273: Schubversuche an Balken mit veränderlicher Trägerhöhe (1977).
Von *F. S. Rostásy, K. Roeder* und *F. Leonhardt.* 9,70 EUR

274: Witterungsbeständigkeit von Beton, 2. Bericht (1977).
Von *K. Walz* und *E. Hartmann.* 8,40 EUR

275: Schubversuche an Balken und Platten bei gleichzeitigem Längszug (1977).
Von *F. Leonhardt, F. S. Rostásy, J. MacGregor* und *M. Patzak.* 11,00 EUR

276: Versuche an zugbeanspruchten Übergreifungsstößen von Rippenstählen (1977).
Von *S. Stöckl, B. Menne* und *H. Kupfer.* 15,50 EUR

277: Versuchsergebnisse zur Festigkeit und Verformung von Beton bei mehraxialer Druckbeanspruchung – Results of Test Concerning Strength and Strain of Concrete Subjected to Multiaxial Compressive Stresses (1977).
Von *G. Schickert* und *H. Winkler.* 17,20 EUR

278: Berechnungen von Temperatur- und Feuchtefeldern in Massivbauten nach der Methode der Finiten Elemente (1977).
Von *J. H. Argyris, E. P. Warnke* und *K. J. Willam.* 10,10 EUR

279: Finite Elementberechnung von Spannbeton-Reaktordruckbehältern.
Von *J. H. Argyris, G. Faust, J. Szimmat, E. P. Warnke* und *K. J. Willam.*
Zur Konvertierung von SMART I (1977).
Von *J. H. Argyris, J. Szimmat* und *K. J. Willam.* 11,50 EUR

Heft

280: Nichtisothermer Feuchtetransport in dickwandigen Betonteilen von Reaktordruckbehältern.
Von *K. Kiessl* und *K. Gertis.*
Zur Wärme- und Feuchtigkeitsleitung in Beton.
Von *J. Hundt.*
Einfluss des Wassergehalts auf die Eigenschaften des erhärteten Betons (1977).
Von *M. J. Setzer.* 14,40 EUR

281: Untersuchungen über das Verhalten von Beton bei schlagartiger Beanspruchung (1977).
Von *C. Popp.* 7,90 EUR

282: Vorausbestimmung der Spannkraftverluste infolge Dehnungsbehinderung (1977).
Von *R. Walther, U. Utescher* und *D. Schreck.* 8,90 EUR

283: Technische Möglichkeiten zur Erhöhung der Zugfestigkeit von Beton (1977).
Von *G. Rehm, P. Diem* und *R. Zimbelmann.* 13,10 EUR

284: Experimentelle und theoretische Untersuchungen zur Lasteintragung in die Bewehrung von Stahlbetondruckgliedern (1977).
Von *F. P. Müller* und *W. Eisenbiegler.* 8,20 EUR

285: Zur Traglast der ausmittig gedrückten Stahlbetonstütze mit Umschnürungsbewehrung (1977).
Von *B. Menne.* 8,60 EUR

286: Versuche über Teilflächenbelastung von Normalbeton (1977).
Von *P. Wurm* und *F. Daschner.* 10,70 EUR

287: Spannbetonbehälter für Siedewasserreaktoren mit einer Leistung von 1600 MWe (1977).
Von *F. Bremer* und *W. Spandick.* 6,80 EUR

288: Tragverhalten von aus Fertigteilen zusammengesetzten Scheiben.
Von *G. Mehlhorn* und *H. Schwing.*
Versuche zur Schubtragfähigkeit verzahnter Fugen (1977).
Von *G. Mehlhorn, H. Schwing* und *K.-R. Berg.* vergriffen

289: Prüfverfahren zur Beurteilung von Rostschutzmitteln für die Bewehrung von Gasbeton.
Von *W. Manns, H. Schneider, R. Schönfelder.*
Frostwiderstand von Beton.
Von *W. Manns* und *E. Hartmann.*
Zum Einfluss von Mineralölen auf die Festigkeit von Beton (1977).
Von *W. Manns* und *E. Hartmann.* 8,60 EUR

290: Studie über den Abbruch von Spannbeton-Reaktordruckbehältern.
Von *K. Kleiser, K. Essig, K. Cerff* und *H. K. Hilsdorf.*
Grundlagen eines Modells zur Beschreibung charakteristischer Eigenschaften des Betons (1977).
Von *F. H. Wittmann.* 14,40 EUR

291: Übergreifungsstöße von Rippenstäben unter schwellender Belastung.
Von *G. Rehm* und *R. Eligehausen.*
Übergreifungsstöße geschweißter Betonstahlmatten (1977).
Von *G. Rehm, R. Tewes* und *R. Eligehausen.* 10,70 EUR

Heft

292: Lösung versuchstechnischer Fragen bei der Ermittlung des Festigkeits- und Verformungsverhaltens von Beton unter dreiachsiger Belastung (1978).
Von *D. Linse.* 8,40 EUR

293: Zur Messtechnik für die Sicherheitsbeurteilung und -überwachung von Spannbeton-Reaktordruckbehältern (1978).
Von *N. Czaika, N. Mayer, C. Amberg, G. Magiera, G. Andreae* und *W. Markowski.* 11,50 EUR

294: Studien zur Auslegung von Spannbetondruckbehältern für wassergekühlte Reaktoren (1978).
Von *K. Schimmelpfennig, G. Bäätjer, U. Eckstein, U. Ick* und *S. Wrage.* 10,70 EUR

295: Kriech- und Relaxationsversuche an sehr altem Beton.
Von *H. Trost, H. Cordes* und *G. Abele.*
Kriechen und Rückkriechen von Beton nach langer Lasteinwirkung.
Von *P. Probst und S. Stöckl.*
Versuche zum Einfluss des Belastungsalters auf das Kriechen von Beton (1978).
Von *K. Wesche, I. Schrage* und *W. vom Berg.* 14,40 EUR

296: Die Bewehrung von Stahlbetonbauteilen bei Zwangsbeanspruchung infolge Temperatur (1978).
Von *P. Noakowski.* vergriffen

297: Einfluss des Feuchtigkeitsgehaltes und des Reifegrades auf die Wärmeleitfähigkeit von Beton.
Von *J. Hundt* und *A. Wagner.*
Sorptionsuntersuchungen am Zementstein, Zementmörtel und Beton (1978).
Von *J. Hundt* und *H. Kantelberg.* 8,60 EUR

298: Erfahrungen bei der Prüfung von temporären Korrosionsschutzmitteln für Spannstähle.
Von *G. Rieche* und *J. Delille.*
Untersuchungen über den Korrosionsschutz von Spannstählen unter Spritzbeton (1978).
Von *G. Rehm, U. Nürnberger* und *R. Zimbelmann.* 8,10 EUR

299: Versuche an dickwandigen, unbewehrten Betonringen mit Innendruckbeanspruchung (1978).
Von *J. Neuner, S. Stöckl* und *E. Grasser.* 8,60 EUR

300: Hinweise zu DIN 1045, Ausgabe Dezember 1978. Bearbeitet von *D. Bertram* und *H. Deutschmann.*
Erläuterung der Bewehrungsrichtlinien (1979).
Von *G. Rehm, R. Eligehausen* und *B. Neubert.* vergriffen

301: Übergreifungsstöße zugbeanspruchter Rippenstäbe mit geraden Stabenden (1979).
Von *R. Eligehausen.* 12,90 EUR

302: Einfluss von Zusatzmitteln auf den Widerstand von jungem Beton gegen Rissbildung bei scharfem Austrocknen.
Von *W. Manns* und *K. Zeus.*
Spannungsoptische Untersuchungen zum Tragverhalten von zugbeanspruchten Übergreifungsstößen (1979).
Von *M. Betzle.* 8,60 EUR

303: Querkraftschlüssige Verbindung von Stahlbetondeckenplatten (1979).
Von *H. Paschen* und *V. C. Zillich.* 10,70 EUR

304: Kunstharzgebundene Glasfaserstäbe als Bewehrung im Betonbau.
Von *G. Rehm* und *L. Franke.*
Zur Frage der Krafteinleitung in kunstharzgebundene Glasfaserstäbe (1979).
Von *G. Rehm, L. Franke* und *M. Patzak.* 9,40 EUR

305: Vorherbestimmung und Kontrolle des thermischen Ausdehnungskoeffizienten von Beton (1979).
Von *S. Ziegeldorf K. Kleiser* und *H. K. Hilsdorf.* 7,30 EUR

306: Dreidimensionale Berechnung eines Spannbetonbehälters mit heißer Dichthaut für einen 1500 MWe Druckwasserreaktor (1979).
Von *E. Ettel, H. Hinterleitner, J. Német, A. Jungmann* und *H. Kopp.* 8,10 EUR

307: Zur Bemessung der Schubbewehrung von Stahlbetonbalken mit möglichst gleichmäßiger Zuverlässigkeit (1979).
Von *W. Moosecker.* 8,10 EUR

308: Tragfähigkeit auf schrägen Druck von Brückenstegen, die durch Hüllrohre geschwächt sind.
Von *R. Koch* und *F. S. Rostásy.*
Spannungszustand aus Vorspannung im Bereich gekrümmter Spannglieder (1979).
Von *V. Cornelius* und *G. Mehlhorn.* 10,10 EUR

309: Kunstharzmörtel und Kunstharzbetone unter Kurzzeit- und Dauerstandbelastung.
Von *G. Rehm, L. Franke* und *K. Zeus.*
Langzeituntersuchungen an epoxidharzverklebten Zementmörtelprismen (1980).
Von *P. Jagfeld.* 10,00 EUR

310: Teilweise Vorspannung – Verbundfestigkeit von Spanngliedern und ihre Bedeutung für Rissbildung und Rissbreitenbeschränkung (1980).
Von *H. Trost, H. Cordes, U. Thormaehlen* und *H. Hagen.* 19,90 EUR

311: Segmentäre Spannbetonträger im Brückenbau (1980).
Von *K. Guckenberger, F. Daschner* und *H. Kupfer.* 18,00 EUR

312: Schwellenwerte beim Betondruckversuch (1980).
Von *G. Schickert.* 18,00 EUR

313: Spannungs-Dehnungs-Linien von Leichtbeton.
Von *H. Herrmann.*
Versuche zum Kriechen und Schwinden von hochfestem Leichtbeton (1980).
Von *P. Probst* und *S. Stöckl.* 14,50 EUR

314: Kurzzeitverhalten von extrem leichten Betonen, Druckfestigkeit und Formänderungen.
Von *K. Bastgen* und *K. Wesche.*
Die Schubtragfähigkeit bewehrter Platten und Balken aus dampfgehärtetem Gasbeton nach Versuchen (1980).
Von *D. Briesemann.* 22,30 EUR

315: Bestimmung der Beulsicherheit von Schalen aus Stahlbeton unter Berücksichtigung der physikalisch-nicht-linearen Materialeigenschaften (1980).
Von *W. Zerna, I. Mungan* und *W. Steffen.* 7,60 EUR

316: Versuche zur Bestimmung der Tragfähigkeit stumpf gestoßener Stahlbetonfertigteilstützen (1980).
Von *H. Paschen* und *V. C. Zillich.* vergriffen

317: Untersuchungen über die Schwingfestigkeit geschweißter Betonstahlverbindungen (1981).
Teil 1: Schwingfestigkeitsversuche.
Von *G. Rehm, W. Harre* und *D. Russwurm.*
Teil 2: Werkstoffkundliche Untersuchungen.
Von *G. Rehm* und *U. Nürnberger.* 17,20 EUR

318: Eigenschaften von feuerverzinkten Überzügen auf kaltumgeformten Betonrippenstählen und Betonstahlmatten aus kaltgewälztem Betonrippenstahl.
Technologische Eigenschaften von kaltgeformten Betonrippenstählen und Betonstahlmatten aus kaltgewalztem Betonrippenstahl nach einer Feuerverzinkung (1981).
Von *U. Nürnberger.* 9,40 EUR

319: Vollstöße durch Übergreifung von zugbeanspruchten Rippenstählen in Normalbeton.
Von *M. Betzle, S. Stöckl* und *H. Kupfer.*
Vollstöße durch Übergreifung von zugbeanspruchten Rippenstählen in Leichtbeton.
Von *S. Stöckl, M. Betzle* und *G. Schmidt-Thrö.*
Verbundverhalten von Betonstählen, Untersuchung auf der Grundlage von Ausziehversuchen.
Von *H. Martin* und *P. Noakowski.*
Ermittlung der Verbundspannungen an gedrückten einbetonierten Betonstählen (1981).
Von *F. P. Müller* und *W. Eisenbiegler.* 25,20 EUR

320: Erläuterungen zu DIN 4227 Spannbeton.
Teil 1: Bauteile aus Normalbeton mit beschränkter oder voller Vorspannung, Ausgabe 07.88
Teil 2: Bauteile mit teilweiser Vorspannung, Ausgabe 05.84
Teil 3: Bauteile in Segmentbauart; Bemessung und Ausführung der Fugen, Ausgabe 12.83
Teil 4: Bauteile aus Spannleichtbeton, Ausgabe 02.86
Teil 5: Einpressen von Zementmörtel in Spannkanäle, Ausgabe 12.79
Teil 6: Bauteile mit Vorspannung ohne Verbund, Ausgabe 05.82 (1989).
Zusammengestellt von *D. Bertram.* 34,30 EUR

321: Leichtzuschlag-Beton mit hohem Gehalt an Mörtelporen (1981).
Von *H. Weigler, S. Karl* und *C. Jaegermann.* 6,20 EUR

322: Biegebemessung von Stahlleichtbeton, Ableitung der Spannungsverteilung in der Biegedruckzone aus Prismenversuchen als Grundlage für DIN 4219.
Von *E. Grasser* und *P. Probst.*
Versuche zur Aufnahme der Umlenkkräfte von gekrümmten Bewehrungsstäben durch Betondeckung und Bügel (1981).
Von *J. Neuner* und *S. Stöckl.* 14,50 EUR

323: Zum Schubtragverhalten stabförmiger Stahlbetonelemente (1981).
Von *R. Mallée.* 10,70 EUR

Heft

324: Wärmeausdehnung, Elastizitätsmodul, Schwinden, Kriechen und Restfestigkeit von Reaktorbeton unter einachsiger Belastung und erhöhten Temperaturen.
Von *H. Aschl* und *S. Stöckl*.
Versuche zum Einfluss der Belastungshöhe auf das Kriechen des Betons (1981).
Von *S. Stöckl*. 15,90 EUR

325: Großmodellversuche zur Spanngliedreibung (1981).
Von *H. Cordes*, *K. Schütt* und *H. Trost*. 10,70 EUR

326: Blockfundamente für Stahlbetonfertigstützen (1981).
Von *H. Dieterle* und *A. Steinle*.
vergriffen

327: Versuche zur Knicksicherung von druckbeanspruchten Bewehrungsstäben (1981).
Von *J. Neuner* und *S. Stöckl*.
8,60 EUR

328: Zum Tragfähigkeitsnachweis für Wand-Decken-Knoten im Großtafelbau (1982).
Von *E. Hasse*. 14,50 EUR

329: Sachstandbericht Massenbeton.
Von *Deutscher Beton-Verein e.V.*
Untersuchungen an einem über 20 Jahre alten Spannbetonträger der Pliensaubrücke Esslingen am Neckar (1982).
Von *K. Schäfer* und *H. Scheef*.
8,60 EUR

330: Zusammenstellung und Beurteilung von Messverfahren zur Ermittlung der Beanspruchungen in Stahlbetonbauteilen (1982).
Von *H. Twelmeier* und *J. Schneefuß*.
12,10 EUR

331: Kleben im konstruktiven Betonbau (1982).
Von *G. Rehm* und *L. Franke*.
12,40 EUR

332: Anwendungsgrenzen von vereinfachten Bemessungsverfahren für schlanke, zweiachsig ausmittig beanspruchte Stahlbetondruckglieder.
Von *P. C. Olsen* und *U. Quast*.
Traglast von Druckgliedern mit vereinfachter Bügelbewehrung unter Feuerangriff.
Von *A. Haksever* und *R. Hass*.
Traglast von Druckgliedern mit vereinfachter Bügelbewehrung unter Normaltemperatur und Kurzzeitbeanspruchung (1982).
Von *K. Kordina* und *R. Mester*.
15,00 EUR

333. Festschrift „75 Jahre Deutscher Ausschuß für Stahlbeton" (1982).
Von *D. Bertram*, *E. Bornemann*, *N. Bunke*, *H. Goffin*, *D. Jungwirth*, *K. Kordina*, *H. Kupfer*, *J. Schlaich*, *B. Wedler†* und *W. Zerna*.
22,60 EUR

334: Versuche an Spannbetonbalken unter kombinierter Beanspruchung aus Biegung, Querkraft und Torsion (1982).
Von *M. Teutsch* und *K. Kordina*.
10,20 EUR

335: Versuche zum Tragverhalten von segmentären Spannbetonträgern – Vergleichende Auswertung für Epoxidharz- und Zementmörtelfugen (1982).
Von *H. Kupfer*, *K. Guckenberger* und *F. Daschner*. 10,70 EUR

336: Tragfähigkeit und Verformung von Stahlbetonbalken unter Biegung und gleichzeitigem Zwang infolge Auflagerverschiebung (1982).
Von *K. Kordina*, *F. S. Rostásy* und *B. Svensvik*. 10,70 EUR

337: Verhalten von Beton bei hohen Temperaturen – Behaviour of Concrete at High Temperatures (1982).
Von *U. Schneider*. 15,50 EUR

338: Berechnung des zeitabhängigen Verhaltens von Stahlbetonplatten unter Last- und Zwangsbeanspruchung im ungerissenen und gerissenen Zustand (1982).
Von *G. Schaper*. 13,40 EUR

339: Stützenstöße im Stahlbeton-Fertigteilbau mit unbewehrten Elastomerlagern (1982).
Von *F. Müller*, *H. R. Sasse* und *U. Thormählen*. vergriffen

340: Durchlaufende Deckenkonstruktionen aus Spannbetonfertigteilplatten mit ergänzender Ortbetonschicht – Continuous Skin Stressed Slabs (1982).
Behaviour in Bending (Biegetrageverhalten).
Von *J. Rosenthal* und *E. Bljuger*.
Schubtragverhalten (Behaviour in Shear).
Von *F. Daschner* und *H. Kupfer*.
11,60 EUR

341: Zum Ansatz der Betonzugfestigkeit bei den Nachweisen zur Trag- und Gebrauchsfähigkeit von unbewehrten und bewehrten Betonbauteilen (1983).
Von *M. Jahn*. 8,60 EUR

342: Dynamische Probleme im Stahlbetonbau –
Teil I: Der Baustoff Stahlbeton unter dynamischer Beanspruchung (1983).
Von *F. P. Müller†*, *E. Keintzel* und *H. Charlier*. 18,80 EUR

343: Versuche zum Kriechen und Schwinden von hochfestem Leichtbeton. Versuche zum Rückkriechen von hochfestem Leichtbeton (1983).
Von *P. Hofmann* und *S. Stöckl*.
8,10 EUR

344: Versuche zur Teilflächenbelastung von Leichtbeton für tragende Konstruktionen.
Von *H. G. Heilmann*.
Teilflächenbelastung von Normalbeton – Versuche an bewehrten Scheiben (1983).
Von *P. Wurm* und *F. Daschner*.
12,60 EUR

345: Experimentelle Ermittlung der Steifigkeiten von Stahlbetonplatten (1983).
Von *H. Schäfer*, *K. Schneider* und *H. G. Schäfer*. 11,60 EUR

346: Tragfähigkeit geschweißter Verbindungen im Betonfertigteilbau.
Von *E. Cziesielski* und *M. Friedmann*.
Versuche zur Ermittlung der Tragfähigkeit in Beton eingespannter Rundstahldollen aus nichtrostendem austenitischem Stahl.
Von *G. Utescher* und *H. Herrmann*.
Untersuchungen über in Beton eingelassene Scherbolzen aus Betonstahl (1983).
Von *H. Paschen* und *T. Schönhoff*.
vergriffen

347: Wirkung der Endhaken bei Vollstößen durch Übergreifung von zugbeanspruchten Rippenstählen.
Von *G. Schmidt-Thrö*, *S. Stöckl* und *M. Betzle*
Übergreifungs-Halbstoß mit kurzem Längsversatz ($l_v = 0{,}5\ l_ü$) bei zugbeanspruchten Rippenstählen in Leichtbeton.
Von *M. Betzle*, *S. Stöckl* und *H. Kupfer*.
Rissflächen im Beton im Bereich von Übergreifungsstößen zugbeanspruchter Rippenstähle (1983).
Von *M. Betzle*, *S. Stöckl* und *H. Kupfer*. 17,40 EUR

348: Tragfähigkeit querkraftschlüssiger Fugen zwischen Stahlbeton-Fertigteildeckenelementen (1983).
Von *H. Paschen* und *V. C. Zillich*.
vergriffen

349: Bestimmung des Wasserzementwertes von Frischbeton (1984).
Von *H. K. Hilsdorf*. 10,70 EUR

350: Spannbetonbauteile in Segmentbauart unter kombinierter Beanspruchung aus Torsion, Biegung und Querkraft.
Von *K. Kordina*, *M. Teutsch* und *V. Weber*.
Rissbildung von Segmentbauteilen in Abhängigkeit von Querschnittsausbildung und Spannstahlverbundeigenschaften.
Von *K. Kordina* und *V. Weber*.
Einfluss der Ausbildung unbewehrter Pressfugen auf die Tragfähigkeit von schrägen Druckstreben in den Stegen von Segmentbauteilen (1984).
Von *K. Kordina* und *V. Weber*.
16,70 EUR

351: Belastungs- und Korrosionsversuche an teilweise vorgespannten Balken.
Von *Günter Schelling* und *Ferdinand S. Rostásy*.
Teilweise Vorspannung – Plattenversuche (1984).
Von *Kassian Janovic* und *Herbert Kupfer*. 23,90 EUR

352: Empfehlungen für brandschutztechnisch richtiges Konstruieren von Betonbauwerken.
Von *K. Kordina* und *L. Krampf*.
Möglichkeiten, nachträglich die in einem Betonbauteil während eines Schadenfeuers aufgetretenen Temperaturen abzuschätzen.
Von *A. Haksever* und *L. Krampf*.
Brandverhalten von Decken aus Glasstahlbeton nach DIN 1045 (Ausg. 12.78), Abschn. 20.3.
Von *C. Meyer-Ottens*.
Eindringen von Chlorid-Ionen aus PVC-Abbrand in Stahlbetonbauteile – Literaturauswertung (1984).
Von *K. Wesche*, *G. Neroth* und *J. W. Weber*. vergriffen

353: Einpressmörtel mit langer Verarbeitungszeit.
Von *W. Manns* und *R. Zimbelmann*.
Auswirkung von Fehlstellen im Einpressmörtel auf die Korrosion des Spannstahls.
Von *G. Rehm*, *R. Frey* und *D. Funk*.
Korrosionsverhalten verzinkter Spannstähle in gerissenem Beton (1984).
Von *U. Nürnberger*. 30,60 EUR

354: Bewehrungsführung in Ecken und Rahmenendknoten.
Von *Karl Kordina*.
Vorschläge zur Bemessung rechteckiger und kranzförmiger Konsolen insbesondere unter exzentrischer Belastung aufgrund neuer Versuche (1984).
Von *Heinrich Paschen* und *Hermann Malonn*. vergriffen

Heft

355: Untersuchungen zur Vorspannung ohne Verbund.
Von *Heinrich Trost, Heiner Cordes* und *Bernhard Weller.*
Anwendung der Vorspannung ohne Verbund.
Von *Karl Kordina, Josef Hegger* und *Manfred Teutsch.*
Ermittlung der wirtschaftlichen Bewehrung von Flachdecken mit Vorspannung ohne Verbund (1984).
Von *Karl Kordina, Manfred Teutsch* und *Josef Hegger.* 20,90 EUR

356: Korrosionsschutz von Bauwerken, die im Gleitschalungsbau errichtet wurden (1984).
Von *Karl Kordina* und *Siegfried Droese.* 16,70 EUR

357: Konstruktion, Bemessung und Sicherheit gegen Durchstanzen von balkenlosen Stahlbetondecken im Bereich der Innenstützen (1984).
Von *Udo Schaefers.* vergriffen

358: Kriechen von Beton unter hoher zentrischer und exzentrischer Druckbeanspruchung (1985).
Von *Emil Grasser* und *Udo Kraemer.* 15,30 EUR

359: Versuche zur Ermüdungsbeanspruchung der Schubbewehrung von Stahlbetonträgern.
Von *Klaus Guckenberger, Herbert Kupfer* und *Ferdinand Daschner.*
Vorgespannte Schubbewehrung (1985).
Von *Jürgen Ruhnau* und *Herbert Kupfer.* 25,20 EUR

360: Festigkeitsverhalten und Strukturveränderungen von Beton bei Temperaturbeanspruchung bis 250 °C (1985).
Von *Jürgen Seeberger, Jörg Kropp* und *Hubert K. Hilsdorf.* 18,80 EUR

361: Beitrag zur Bemessung von schlanken Stahlbetonstützen für schiefe Biegung mit Achsdruck unter Kurzzeit- und Dauerbelastung – Contribution to the Design of Slender Reinforced Concrete Columns Subjected to Biaxial Bending and Axial Compression Considering Short and Long Term Loadings (1985).
Von *Nelson Szilard Galgoul.* 21,50 EUR

362: Versuche an Konstruktionsleichtbetonbauteilen unter kombinierter Beanspruchung aus Torsion, Biegung und Querkraft (1985).
Von *Karl Kordina* und *Manfred Teutsch.* 13,40 EUR

363: Versuche zur Mitwirkung des Betons in der Zugzone von Stahlbetonröhren (1985).
Von *Jörg Schlaich* und *Hans Schober.* 14,50 EUR

364: Empirische Zusammenhänge zur Ermittlung der Schubtragfähigkeit stabförmiger Stahlbetonelemente (1985).
Von *Karl Kordina* und *Franz Blume.* 11,80 EUR

365: Experimentelle Untersuchungen bewehrter und hohler Prüfkörper aus Normalbeton mittels eines zwängungsarmen Krafteinleitungssystems (1985).
Von *Manfred Specht, Rita Schmidt* und *Hartmut Kappes.* 16,10 EUR

366: Grundsätzliche Untersuchungen zum Geräteeinfluss bei der mehraxialen Druckprüfung von Beton (1985).
Von *Helmut Winkler.* 29,00 EUR

Heft

367: Verbundverhalten von Bewehrungsstählen unter Dauerbelastung in Normal- und Leichtbeton.
Von *Kassian Janovic.*
Übergreifungsstöße geschweißter Betonstahlmatten.
Von *Gallus Rehm* und *Rüdiger Tewes.*
Übergreifungsstöße geschweißter Betonstahlmatten in Stahlleichtbeton (1986).
Von *Gallus Rehm* und *Rüdiger Tewes.* 14,50 EUR

368: Fugen und Aussteifungen in Stahlbetonskelettbauten (1986).
Von *Bernd Hock, Kurt Schäfer* und *Jörg Schlaich.* vergriffen

369: Versuche zum Verhalten unterschiedlicher Stahlsorten in stoßbeanspruchten Platten (1986).
Von *Josef Eibl* und *Klaus Kreuser.* 13,40 EUR

370: Einfluss von Rissen auf die Dauerhaftigkeit von Stahlbeton- und Spannbetonbauteilen.
Von *Peter Schießl.*
Dauerhaftigkeit von Spanngliedern unter zyklischen Beanspruchungen.
Von *Heiner Cordes.*
Beurteilung der Betriebsfestigkeit von Spannbetonbrücken im Koppelfugenbereich unter besonderer Berücksichtigung einer möglichen Rissbildung.
Von *Gert König* und *Hans-Christian Gerhardt.*
Nachweis zur Beschränkung der Rissbreite in den Normen des Deutschen Ausschusses für Stahlbeton (1986).
Von *Eilhard Wölfel.* vergriffen

371: Tragfähigkeit durchstanzgefährdeter Stahlbetonplatten-Entwicklung von Bemessungsvorschlägen (1986).
Von *Karl Kordina* und *Diedrich Nölting.* vergriffen

372: Literaturstudie zur Schubsicherung bei nachträglich ergänzten Querschnitten.
Von *Ferdinand Daschner* und *Herbert Kupfer.*
Versuche zur notwendigen Schubbewehrung zwischen Betonfertigteilen und Ortbeton.
Von *Ferdinand Daschner.*
Verminderte Schubdeckung in Stahlbeton- und Spannbetonträgern mit Fugen parallel zur Tragrichtung unter Berücksichtigung nicht vorwiegend ruhender Lasten.
Von *Ingo Nissen, Ferdinand Daschner* und *Herbert Kupfer.*
Literaturstudie über Versuche mit sehr hohen Schubspannungen (1986).
Von *Herbert Kupfer* und *Ferdinand Daschner.* vergriffen

373: Empfehlungen für die Bewehrungsführung in Rahmenecken und -knoten.
Von *Karl Kordina, Ehrenfried Schaaff* und *Thomas Westphal.*
Das Übertragungs- und Weggrößenverfahren für ebene Stahlbetonstabtragwerke unter Verwendung von Tangentensteifigkeiten (1986).
Von *Poul Colberg Olsen.* vergriffen

374: Schwingfestigkeitsverhalten von Betonstählen unter wirklichkeitsnahen Beanspruchungs- und Umgebungsbedingungen (1986).
Von *Gallus Rehm, Wolfgang Harre* und *Willibald Beul.* 14,50 EUR

Heft

375: Grundlagen und Verfahren für den Knicksicherheitsnachweis von Druckgliedern aus Konstruktionsleichtbeton.
Von *Roland Molzahn.*
Einfluss des Kriechens auf Ausbiegung und Tragfähigkeit schlanker Stützen aus Konstruktionsleichtbeton (1986).
Von *Roland Molzahn.* 13,40 EUR

376: Trag- und Verformungsfähigkeit von Stützen bei großen Zwangsverschiebungen der Decken.
Von *Peter Steidle* und *Kurt Schäfer.*
Versuche an Stützen mit Normalkraft und Zwangsverschiebungen (1986).
Von *Rolf Wohlfahrt* und *Rainer Koch.* 22,60 EUR

377: Versuche zur Schubtragwirkung von profilierten Stahlbeton- und Spannbetonträgern mit überdrückten Gurtplatten (1986).
Von *Herbert Kupfer* und *Klaus Guckenberger.* 14,00 EUR

378: Versuche über das Verbundverhalten von Rippenstählen bei Anwendung des Gleitbauverfahrens.
Teilbericht I:
Ausziehversuche, Proben in Utting hergestellt.
Von *Gerfried Schmidt-Thrö* und *Siegfried Stöckl.*
Teilbericht II:
Versuche zur Bestimmung charakteristischer Betoneigenschaften bei Anwendung des Gleitbauverfahrens.
Von *Gerfried Schmidt-Thrö, Siegfried Stöckl* und *Herbert Kupfer.*
Teilbericht III:
Ausziehversuche und Versuche an Übergreifungsstößen, Proben in Berlin bzw. Köln hergestellt.
Von *Klaus Kluge, Gerfried Schmidt-Thrö, Siegfried Stöckl* und *Herbert Kupfer.*
Einfluss der Probekörperform und der Messpunktanordnung auf die Ergebnisse von Ausziehversuchen (1986).
Von *Gerfried Schmidt-Thrö, Siegfried Stöckl* und *Herbert Kupfer.* 27,40 EUR

379: Experimentelle und analytische Untersuchungen zur wirklichkeitsnahen Bestimmung der Bruchschnittgrößen unbewehrter Betonbauteile unter Zugbeanspruchung, (1987).
Von *Dietmar Scheidler.* 16,70 EUR

380: Eigenspannungszustand in Stahl- und Spannbetonkörpern infolge unterschiedlichen thermischen Dehnverhaltens von Beton und Stahl bei tiefen Temperaturen.
Von *Ferdinand S. Rostásy* und *Jochen Scheuermann.*
Verbundverhalten einbetonierten Betonrippenstahls bei extrem tiefer Temperatur.
Von *Ferdinand S. Rostásy* und *Jochen Scheuermann.*
Versuche zur Biegetragfähigkeit von Stahlbetonplattenstreifen bei extrem tiefer Temperatur (1987).
Von *Günter Wiedemann, Jochen Scheuermann, Karl Kordina* und *Ferdinand S. Rostásy.* 19,90 EUR

381: Schubtragverhalten von Spannbetonbauteilen mit Vorspannung ohne Verbund.
Von *Karl Kordina* und *Josef Hegger.*
Systematische Auswertung von Schubversuchen an Spannbetonbalken (1987).
Von *Karl Kordina* und *Josef Hegger.* 21,50 EUR

Heft

415: Programmgesteuerte Berechnung beliebiger Massivbauquerschnitte unter zweiachsiger Biegung mit Längskraft (Programm MASQUE) (1990).
Von *Dirk Busjaeger* und *Ulrich Quast*.
31,20 EUR

416: Betonbau beim Umgang mit wassergefährdenden Stoffen – Sachstandsbericht (1991).
Von *Thomas Fehlhaber, Gert König, Siegfried Mängel, Hermann Poll, Hans-Wolf Reinhardt, Carola Reuter, Peter Schießl, Bernd Schnütgen, Gerhard Spanka Friedhelm Stangenberg, Gerd Thielen* und *Johann-Dietrich Wörner*. 37,60 EUR

417: Stahlbeton- und Spannbetonbauteile bei extrem tiefer Temperatur – Versuche und Berechnungsansätze für Lasten und Zwang (1991).
Von *Uwe Pusch* und *Ferdinand S. Rostásy*. 22,60 EUR

418: Warmbehandlung von Beton durch Mikrowellen (1991).
Von *Ulrich Schneider* und *Frank Dumat*. 30,00 EUR

419: Bruchmechanisches Verhalten von Beton unter monotoner und zyklischer Zugbeanspruchung (1991).
Von *Herbert Duda*. 17,20 EUR

420: Versuche zum Kriechen und zur Restfestigkeit von Beton bei mehrachsiger Beanspruchung.
Von *Norbert Lanig, Siegfried Stöckl* und *Herbert Kupfer*.
Kriechen von Beton nach langer Lasteinwirkung.
Von *Norbert Lanig* und *Siegfried Stöckl*.
Frühe Kriechverformungen des Betons (1991).
Von *Heinrich Trost* und *Hans Paschmann*. 24,70 EUR

421: Entwicklung radiographischer Untersuchungsmethoden des Verbundverhaltens von Stahl und Beton (1991).
Von *Andrea Steinwedel*.
22,60 EUR

422: Prüfung von Beton-Empfehlungen und Hinweise als Ergänzung zu DIN 1048 (1991).
Zusammengestellt von *Norbert Bunke*.
33,30 EUR

423: Experimentelle Untersuchungen des Trag- und Verformungsverhaltens schlanker Stahlbetondruckglieder mit zweiachsiger Ausmitte.
Von *Rainer Grzeschkowitz, Karl Kordina* und *Manfred Teutsch*.
Erweiterung von Traglastprogrammen für schlanke Stahlbetondruckglieder (1992).
Von *Rainer Grzeschkowitz* und *Ulrich Quast*. 23,60 EUR

424: Tragverhalten von Befestigungen unter Querlasten in ungerissenem Beton (1992).
Von *Werner Fuchs*. 29,00 EUR

425: Bemessungshilfsmittel zu Eurocode 2 Teil 1 (DIN V ENV 1992 Teil 1-1, Ausgabe 06.92).
Planung von Stahlbeton- und Spannbetontragwerken (1992).
3. ergänzte Auflage 1997.
Von *Karl Kordina* u. a. 40,90 EUR

426: Einfluss der Probekörperform auf die Ergebnisse von Ausziehversuchen – Finite-Element-Berechnung – (1992).
Von *Jürgen Mainz* und *Siegfried Stöckl*. 19,30 EUR

Heft

427: Verminderte Schubdeckung in Betonträgern mit Fugen parallel zur Tragrichtung bei sehr hohen Schubspannungen und nicht vorwiegend ruhenden Lasten (1992).
Von *Ferdinand Daschner* und *Herbert Kupfer*. 14,00 EUR

428: Entwicklung eines Expertensystems zur Beurteilung, Beseitigung und Vorbeugung von Oberflächenschäden an Betonbauteilen (1992).
Von *Michael Sohni*. 20,40 EUR

429: Der Einfluss mechanischer Spannungen auf den Korrosionswiderstand zementgebundener Baustoffe (1992).
Von *Ulrich Schneider, Erich Nägele Frank Dumat* und *Steffen Holst*.
20,40 EUR

430: Standardisierte Nachweise von häufigen D-Bereichen (1992).
Von *Mattias Jennewein* und *Kurt Schäfer*. 20,40 EUR

431: Spannungsumlagerungen in Verbundquerschnitten aus Fertigteilen und Ortbeton statisch bestimmter Träger infolge Kriechen und Schwinden unter Berücksichtigung der Rissbildung (1992).
Von *Günther Ackermann, Erich Raue, Lutz Ebel* und *Gerhard Setzpfandt*.
vergriffen

432: Lineare und nichtlineare Theorie des Kriechens und der Relaxation von Beton unter Druckbeanspruchung (1992).
Von *Jing-Hua Shen*. 12,90 EUR

433: Zur chloridinduzierten Makroelementkorrosion von Stahl in Beton (1992).
Von *Michael Raupach*. 23,60 EUR

434: Beurteilung der Wirksamkeit von Steinkohlenflugaschen als Betonzusatzstoff (1993).
Von *Franz Sybertz*. 23,60 EUR

435: Zur Spannungsumlagerung im Spannbeton bei der Rissbildung unter statischer und wiederholter Belastung (1993).
Von *Nguyen Viet Tue*. 18,30 EUR

436: Zum karbonatisierungsbedingten Verlust der Dauerhaftigkeit von Außenbauteilen aus Stahlbeton (1993).
Von *Dieter Bunte*. 27,90 EUR

437: Festigkeit und Verformung von Beton bei hoher Temperatur und biaxialer Beanspruchung – Versuche und Modellbildung – (1994).
Von *Karl-Christian Thienel*.
22,60 EUR

438: Hochfester Beton, Sachstandsbericht, Teil 1: Betontechnologie und Betoneigenschaften.
Von *Ingo Schrage*.
Teil 2: Bemessung und Konstruktion (1994).
Von *Gert König, Harald Bergner, Rainer Grimm, Markus Held, Gerd Remmel* und *Gerd Simsch*.
19,30 EUR

439: Ermüdungsfestigkeit von Stahlbeton und Spannbetonbauteilen mit Erläuterungen zu den Nachweisen gemäß CEB-FIP. Model Code 1990 (1994).
Von *Gert König* und *Ireneusz Danielewicz*. 21,50 EUR

Heft

440: Untersuchung zur Durchlässigkeit von faserfreien und faserverstärkten Betonbauteilen mit Trennrissen.
Von *Masaaki Tsukamoto*.
Gitterschnittkennwert als Kriterium für die Adhäsionsgüte von Oberflächenschutzsystemen auf Beton (1994).
Von *Michael Fiebrich*. 18,30 EUR

441: Physikalisch nichtlineare Berechnung von Stahlbetonplatten im Vergleich zur Bruchlinientheorie (1994).
Von *Andreas Pardey*. 36,50 EUR

442: Versuche zum Kriechen von Beton bei mehrachsiger Beanspruchung – Auswertung auf der Basis von errechneten elastischen Anfangsverformungen.
Von *Henric Bierwirth, Siegfried Stöckl* und *Herbert Kupfer*.
Kriechen, Rückkriechen und Dauerstandfestigkeit von Beton bei unterschiedlichem Feuchtegehalt und Verwendung von Portlandzement bzw. Portlandkalksteinzement (1994).
Von *Dirk Nechvatal, Siegfried Stöckl* und *Herbert Kupfer*. 20,40 EUR

443: Schutz und Instandsetzung von Betonbauteilen unter Verwendung von Kunststoffen – Sachstandsbericht – (1994).
Von *H. Rainer Sasse* u. a.
51,60 EUR

444: Zum Zug- und Schubtragverhalten von Bauteilen aus hochfestem Beton (1994).
Von *Gerd Remmel*. 23,60 EUR

445: Zum Eindringverhalten von Flüssigkeiten und Gasen in ungerissenen Beton.
Von *Thomas Fehlhaber*.
Eindringverhalten von Flüssigkeiten in Beton in Abhängigkeit von der Feuchte der Probekörper und der Temperatur.
Von *Massimo Sosoro* und *Hans-Wolf Reinhardt*.
Untersuchung der Dichtheit von Vakumbeton gegenüber wassergefährdenden Flüssigkeiten (1994).
Von *Reinhard Frey* und *Hans-Wolf Reinhardt*. 27,90 EUR

446: Modell zur Vorhersage des Eindringverhaltens von organischen Flüssigkeiten in Beton (1995).
Von *Massimo Sosoro*. 17,20 EUR

447: Versuche zum Verhalten von Beton unter dreiachsiger Kurzzeitbeanspruchung.
Tests on the Behaviour of Concrete under Triaxial Shorttime Loading.
Von *Ulrich Scholz, Dirk Nechvatal, Helmut Aschl, Diethelm Linse, Emil Grasser* und *Herbert Kupfer*.
Auswertung von Versuchen zur mehrachsigen Betonfestigkeit, die an der Technischen Universität München durchgeführt wurden.
Evaluation of the Multiaxial Strength of Concrete Tested at Technische Universität München.
Von *Zhenhai Guo, Yunlong Zhou* und *Dirk Nechvatal*.
Versuche zur Methode der Verformungsmessung an dreiachsig beanspruchten Betonwürfeln.
Tests on Methods for Strain Measurements on Cubic Specimen of Concrete under Triaxial Loading (1995).
Von *Christian Dialer, Norbert Lanig, Siegfried Stöckl* und *Cölestin Zelger*.
25,70 EUR

Heft

476: Zuverlässigkeit des Verpressens von Spannkanälen unter Berücksichtigung der Unsicherheiten auf der Baustelle (1997).
Von *Ferdinand S. Rostásy* und *Alex-W. Gutsch.* 25,70 EUR

477: Einfluss bruchmechanischer Kenngrößen auf das Biege- und Schubtragverhalten hochfester Betone (1997).
Von *Rainer Grimm.* 27,90 EUR

478: Tragfähigkeit von Druckstreben und Knoten in D-Bereichen (1997).
Von *Wolfgang Sundermann* und *Kurt Schäfer.* 29,00 EUR

479: Über das Brandverhalten punktgestützter Stahlbetonplatten (1997).
Von *Karl Kordina.* 25,70 EUR

480: Versagensmodell für schubschlanke Balken (1997).
Von *Jürgen Fischer.* 19,30 EUR

481: Sicherheitskonzept für Bauten des Umweltschutzes.
Von *Daniela Kiefer.*
Erfahrungen mit Bauten des Umweltschutzes.
Von *Johann-Dietrich Wörner, Daniela Kiefer* und *Hans-Werner Nordhues.*
Qualitätskontrollmaßnahmen bei Betonkonstruktionen (1997).
Von *Otto Kroggel.* 21,50 EUR

482: Rissbreitenbeschränkung zwangbeanspruchter Bauteile aus hochfestem Normalbeton (1997).
Von *Harald Bergner.* 25,70 EUR

483: Durchlässigkeitsgesetze für Flüssigkeiten mit Feinstoffanteilen bei Betonbunkern von Abfallbehandlungsanlagen.
Von *Klaus-Peter Grote.*
Einfluss von Stahlfasern auf die Durchlässigkeit von Beton (1997).
Von *Ralf Winterberg.* 22,60 EUR

484: Grenzen der Anwendung nichtlinearer Rechenverfahren bei Stabtragwerken und einachsig gespannten Platten.
Von *Rolf Eligehausen* und *Eckhart Fabritius.*
Rotationsfähigkeit von plastischen Gelenken im Stahl- und Spannbetonbau.
Von *Longfei Li.*
Verdrehfähigkeit plastizierter Trag--werksbereiche im Stahlbetonbau (1998).
Von *Peter Langer.* 37,60 EUR

485: Verwendung von Bitumen als Gleitschicht im Massivbau.
Von *Manfred Curbach* und *Thomas Bösche.*
Versuche zur Eignung industriell gefertigter Bitumenbahnen als Bitumengleitschicht (1998).
Von *Manfred Curbach* und *Thomas Bösche.* 21,50 EUR

486: Trag- und Verformungsverhalten von Rahmenknoten (1998).
Von *Karl Kordina, Manfred Teutsch* und *Erhard Wegener.* 34,30 EUR

487: Dauerhaftigkeit hochfester Betone (1998).
Von *Ulf Guse* und *Hubert K. Hilsdorf.* 19,30 EUR

488: Sachstandsbericht zum Einsatz von Textilien im Massivbau (1998).
Von *Manfred Curbach* u. a. 22,60 EUR

Heft

489: Mindestbewehrung für verformungsbehinderte Betonbauteile im jungen Alter (1998).
Von *Udo Paas.* 23,60 EUR

490: Beschichtete Bewehrung. Ergebnisse sechsjähriger Auslagerungsversuche.
Von *Klaus Menzel, Frank Schulze* und *Hans-Wolf Reinhardt.*
Kontinuierliche Ultraschallmessung während des Erstarrens und Erhärtens von Beton als Werkzeug des Qualitätsmanagements (1998).
Von *Hans-Wolf Reinhardt, Christian U. Große* und *Alexander Herb.* 18,30 EUR

491: Der Einfluss der freien Schwingungen auf ausgewählte dynamische Parameter von Stahlbetonbiegeträgern (1999).
Von *Manfred Specht* und *Michael Kramp.* 31,20 EUR

492: Nichtlineares Last-Verformungs-Verhalten von Stahlbeton- und Spannbetonbauteilen, Verformungsvermögen und Schnittgrößenermittlung (1999).
Von *Gert König, Dieter Pommerening* und *Nguyen Viet Tue.* 26,90 EUR

493: Leitfaden für die Erfassung und Bewertung der Materialien eines Abbruchobjektes (1999).
Von *Theo Rommel, Wolfgang Katzer, Gerhard Tauchert* und *Jie Huang.* 18,80 EUR

494: Tragverhalten von Stahlfaserbeton (1999).
Von *Yong-zhi Lin.* 23,60 EUR

495: Stoffeigenschaften jungen Betons; Versuche und Modelle (1999).
Von *Alex-W. Gutsch.* 29,50 EUR

496: Entwerfen und Bemessen von Betonbrücken ohne Fugen und Lager (1999).
Von *Stephan Engelsmann, Jörg Schlaich* und *Kurt Schäfer.* 25,70 EUR

497: Entwicklung von Verfahren zur Beurteilung der Kontaminierung der Baustoffe vor dem Abbruch (Schnellprüfverfahren) (2000).
Von *Jochen Stark* und *Peter Nobst.* 20,90 EUR

498: Kriechen von Beton unter Zugbeanspruchung (2000).
Von *Karl Kordina, Lothar Schubert* und *Uwe Troitzsch.* 16,70 EUR

499: Tragverhalten von stumpf gestoßenen Fertigteilstützen aus hochfestem Beton (2000).
Von *Jens Minnert.* 29,00 EUR

500: BiM-Online – Das interaktive Informationssystem zu „Baustoffkreislauf im Massivbau" (2000).
Von *Hans-Wolf Reinhardt, Marcus Schreyer* und *Joachim Schwarte.* 21,50 EUR

501: Tragverhalten und Sicherheit betonstahlbewehrter Stahlfaserbetonbauteile (2000).
Von *Ulrich Gossla.* 20,40 EUR

502: Witterungsbeständigkeit von Beton. 3. Bericht (2000).
Von *Wilhelm Manns* und *Kurt Zeus.* 17,80 EUR

Heft

503: Untersuchungen zum Einfluss der bezogenen Rippenfläche von Bewehrungsstäben auf das Tragverhalten von Stahlbetonbauteilen im Gebrauchs- und Bruchzustand (2000).
Von *Rolf Eligehausen* und *Utz Mayer.* 20,90 EUR

504: Schubtragverhalten von Stahlbetonbauteilen mit rezyklierten Zuschlägen (2000).
Von *Sufang Lü.* 24,70 EUR

505: Biegetragverhalten von Stahlbetonbauteilen mit rezyklierten Zuschlägen (2000).
Von *Matthias Meißner.* 29,00 EUR

506: Verwertung von Brechsand aus Bauschutt (2000).
Von *Christoph Müller* und *Bernd Dora.* 24,70 EUR

507: Betonkennwerte für die Bemessung und Verbundverhalten von Beton mit rezykliertem Zuschlag (2000).
Von *Konrad Zilch* und *Frank Roos.* 19,30 EUR

508: Zulässige Toleranzen für die Abweichungen der mechanischen Kennwerte von Beton mit rezykliertem Zuschlag (2000).
Von *Johann-Dietrich Wörner, Pieter Moerland, Sabine Giebenhain, Harald Kloft* und *Klaus Leiblein.* 16,70 EUR

509: Bruchmechanisches Verhalten jungen Betons (2000).
Von *Karim Hariri.* 24,70 EUR

510: Probabilistische Lebensdauerbemessung von Stahlbetonbauwerken – Zuverlässigkeitsbetrachtungen zur wirksamen Vermeidung von Bewehrungskorrosion (2000).
Von *Christoph Gehlen.* 24,20 EUR

511: Hydroabrasionsverschleiß von Betonoberflächen.
Beton und Mörtel für die Instandsetzung verschleißgeschädigter Betonbauteile im Wasserbau (2000).
Von *Gesa Haroske, Jan Vala* und *Ulrich Diederichs.* 27,40 EUR

512: Zwang und Rissbildung infolge Hydratationswärme – Grundlagen Berechnungsmodelle und Tragverhalten (2000).
Von *Benno Eierle* und *Karl Schikora.* 27,40 EUR

513: Beton als kreislaufgerechter Baustoff (2001).
Von *Christoph Müller.* 65,50 EUR

514: Einfluss von rezykliertem Zuschlag aus Betonbruch auf die Dauerhaftigkeit von Beton.
Von *Beatrix Kerkhoff* und *Eberhard Siebel.*
Einfluss von Feinstoffen aus Betonbruch auf den Hydratationsfortschritt.
Von *Walter Wassing.*
Recycling von Beton, der durch eine Alkalireaktion gefährdet oder bereits geschädigt ist.
Von *Wolfgang Aue.*
Frostwiderstand von rezykliertem Zuschlag aus Altbeton und mineralischen Baustoffgemischen (Bauschutt) (2001).
Von *Stefan Wies* und *Wilhelm Manns.* 48,60 EUR

Heft

515: Analytische und numerische Untersuchungen des Durchstanzverhaltens punktgestützter Stahlbetonplatten (2001).
Von *Markus Anton Staller.*
43,50 EUR

516: Sachstandbericht Selbstverdichtender Beton (SVB) (2001).
Von *Hans-Wolf Reinhardt, Wolfgang Brameshuber, Geraldine Buchenau, Frank Dehn, Horst Grube, Peter Grübl, Bernd Hillemeier, Martin Jooß, Bert Kilanowski, Thomas Krüger, Christoph Lemmer, Viktor Mechterine, Harald Müller, Thomas Müller, Markus Plannerer, Andreas Rogge, Andreas Schaab, Angelika Schießl* und *Stephan Uebachs.* 33,80 EUR

517: Verformungsverhalten und Tragfähigkeit dünner Stege von Stahlbeton- und Spannbetonträgern mit hoher Betongüte (2001).
Von *Karl-Heinz Reineck, Rolf Wohlfahrt* und *Harianto Hardjasaputra.*
54,20 EUR

518: Schubtragfähigkeit längsbewehrter Porenbetonbauteile ohne Schubbewehrung.
Thermische Vorspannung bewehrter Porenbetonbauteile.
Kriechen von unbewehrtem Porenbeton.
Kriechen des Porenbetons im Bereich der zur Verankerung der Längsbewehrung dienenden Querstäbe und Tragfähigkeit der Verankerung (2001).
Von *Ferdinand Daschner* und *Konrad Zilch.* 55,90 EUR

519: Betonbau beim Umgang mit wassergefährdenden Stoffen. Zweiter Sachstandsbericht mit Beispielsammlung (2001).
Von *Rolf Breitenbücher, Franz-Josef Frey, Horst Grube, Wilhelm Kanning, Klaus Lehmann, Hans-Wolf Reinhardt, Bernd Schnütgen, Manfred Teutsch, Günter Timm* und *Johann-Dietrich Wörner.* 52,10 EUR

520: Frühe Risse in massigen Betonbauteilen – Ingenieurmodelle für die Planung von Gegenmaßnahmen (2001).
Von *Ferdinand S. Rostásy* und *Matias Krauß.* 39,20 EUR

521: Sachstandbericht Nachhaltig Bauen mit Beton (2001).
Von *Hans-Wolf Reinhardt, Wolfgang Brameshuber, Carl-Alexander Graubner, Peter Grübl, Bruno Hauer, Katja Hüske, Julian Kümmel, Hans-Ulrich Litzner, Heiko Lünser, Dieter Rußwurm.* 31,10 EUR

522: Anwendung von hochfestem Beton im Brückenbau.
Von *Konrad Zilch* und *Markus Hennecke.*
Erfahrungen mit Entwurf, Ausschreibung, Vergabe und Tragwerksplanung.
Von *André Müller, Hans Pfisterer, Jürgen Weber* und *Konrad Zilch.*
Erfahrungen mit der Bauausführung und Maßnahmen zur Gewährleistung der geforderten Qualität.
Von *Markus Hennecke, Gert Leonhardt* und *Rolf Stahl.*
Betontechnologie (2002).
Von *Volker Hartmann* und *Werner Schrub.* 37,60 EUR

Heft

523: Beständigkeit verschiedener Betonarten im Meerwasser und in sulfathaltigem Wasser (2003).
Von *Ottokar Hallauer.* 96,10 EUR

524: Mehraxiale Festigkeit von duktilem Hochleistungsbeton (2002).
Von *Manfred Curbach* und *Kerstin Speck.* 68,30 EUR

525: Erläuterungen zu DIN 1045-1; 2. überarbeitete Auflage (2010)
64,30 EUR

526: Erläuterungen zu den Normen DIN EN 206-1, DIN 1045-2, DIN 1045-3, DIN 1045-4 und DIN EN 12620; 2. überarbeitete Auflage (2011).
88,40 EUR

527: Füllen von Rissen und Hohlräumen in Betonbauteilen (2006).
Von *Angelika Eßer.* 58,40 EUR

528: Schubtragfähigkeit von Betonergänzungen an nachträglich aufgerauten Betonoberflächen bei Sanierungs- und Ertüchtigungsmaßnahmen (2002).
Von *Konrad Zilch* und *Jürgen Mainz.*
20,80 EUR

529: Betonwaren mit Recyclingzuschlägen.
Von *Christoph Müller* und *Peter Schießl.*
Rezyklieren von Leichtbeton (2002).
Von *Hans-Wolf Reinhardt* und *Julian Kümmel.* 32,20 EUR

530: Nachweise zur Sicherheit beim Abbruch von Stahlbetonbauwerken durch Sprengen.
Von *Josef Eibl, Andreas Plotzitza, Nico Herrmann.*
Sprengtechnischer Abbruch, Erprobung und Optimierung (2000).
Von *Hans-Ulrich Freund, Gerhard Duseberg, Steffen Schumann, Helmut Roller, Walter Werner.*
36,50 EUR

531: Großtechnische Versuche zur Nassaufbereitung von Recycling-Baustoffen mit der Setzmaschine.
Von *Harald Kurkowski* und *Klaus Mesters.*
Einflüsse der Aufbereitung von Bauschutt für eine Verwendung als Betonzuschlag (2003).
Von *Werner Reichel* und *Petra Heldt.*
42,80 EUR

532: Die Bemessung und Konstruktion von Rahmenknoten. Grundlagen und Beispiele gemäß DIN 1045-1(2002).
Von *Josef Hegger* und *Wolfgang Roeser.* 62,80 EUR

533: Rechnerische Untersuchung der Durchbiegung von Stahlbetonplatten unter Ansatz wirklichkeitsnaher Steifigkeiten und Lagerungsbedingungen und unter Berücksichtigung zeitabhängiger Verformungen (2006).
Von *Konrad Zilch* und *Uli Donaubauer.*
Zum Trag- und Verformungsverhalten bewehrter Betonquerschnitte im Grenzzustand der Gebrauchstauglichkeit.
Von *Wolfgang Krüger* und *Olaf Mertzsch.*
67,70 EUR

534: Sicherheitskonzept für nichtlineare Traglastverfahren im Betonbau (2003).
Von *Michael Six.* 51,90 EUR

535: Rotationsfähigkeit von Rahmenecken (2002).
Von *Jan Akkermann* und *Josef Eibl.*
43,70 EUR

Heft

537: Zum Einfluss der Oberflächengestalt von Rippenstählen auf das Trag- und Verformungsverhalten von Stahlbetonbauteilen (2003).
Von *Utz Mayer.* 44,20 EUR

538: Analyse der Transportmechanismen für wassergefährdende Flüssigkeiten in Beton zur Berechnung des Medientransportes in ungerissene und gerissene Betondruckzonen (2002).
Von *Norbert Brauer.* 45,40 EUR

539: Alkalireaktion im Bauwerksbeton. Ein Erfahrungsbericht (2003).
Von *Wilfried Bödeker.*
26,30 EUR

540: Trag- und Verformungsverhalten von Stahlbetontragwerken unter Betriebsbelastung (2003).
Von *Thomas M. Sippel.*
27,30 EUR

541: Das Ermüdungsverhalten von Dübelbefestigungen (2003).
Von *Klaus Block und Friedrich Dreier.*
38,80 EUR

542: Charakterisierung, Modellierung und Bewertung des Auslaugverhaltens umweltrelevanter, anorganischer Stoffe aus zementgebundenen Baustoffen (2003).
Von *Inga Hohberg.* 52,40 EUR

543: Mikrostrukturuntersuchungen zum Sulfatangriff bei Beton (2003).
Von *Winfried Malorny.* 19,60 EUR

544: Hochfester Beton unter Dauerzuglast (2003).
Von *Tassilo Rinder.* 37,70 EUR

545: Gebrauchsverhalten von Bodenplatten aus Beton unter Einwirkungen infolge Last und Zwang (2004).
Von *Peter Niemann.* 65,00 EUR

546: Zu Deckenscheiben zusammengespannte Stahlbetonfertigteile für demontable Gebäude (2003).
Von *Georg Christian Weiß.*
39,90 EUR

547: Durchstanzen von Bodenplatten unter rotationssymmetrischer Belastung (2004).
Von *Maike Timm.* 49,10 EUR

548: Die Druckfestigkeit von gerissenen Scheiben aus Hochleistungsbeton und selbstverdichtendem Beton unter Berücksichtigung des Einflusses der Rissneigung (2005).
Von *Angelika Schießl.* 56,30 EUR

549: Zum Gebrauchs- und Tragverhalten von Tunnelschalen aus Stahlfaserbeton und stahlfaserverstärktem Stahlbeton (2004).
Von *Olaf Hemmy.* 74,20 EUR

550: Zur Querkrafttragfähigkeit von Balken aus stahlfaserverstärktem Stahlbeton (2004).
Von *Joachim Rosenbusch.*
47,60 EUR

551: Zur Wirkung von Steinkohlenflugasche auf die chloridinduzierte Korrosion von Stahl in Beton (2005).
Von *Udo Wiens.* 63,30 EUR

552: Randbedingungen bei der Instandsetzung nach dem Schutzprinzip W bei Bewehrungskorrosion im karbonatisierten Beton (2005).
Von *Romain Weydert.* 38,50 EUR

Heft

589: Zerstörungsfreie Ortung von Gefügestörungen in Betonbodenplatten (2010).
Von *Harald S. Müller, Martin Fenchel, Herbert Wiggenhauser, Christiane Maierhofer, Martin Krause, Andre Gardei, Frank Mielentz, Boris Milman, Mathias Röllig, Jens Wöstmann.* 84,60 EUR

590: Materialverhalten von hochfestem Beton unter thermomechanischer Beanspruchung (2010).
Von *Sven Huismann.* 65,00 EUR

591: Sachstandbericht Verstärken von Betonbauteilen mit geklebter Bewehrung (2011).
Von *Konrad Zilch, Roland Niedermeier, Wolfgang Finckh.* 77,50 EUR

592: Praxisgerechte Bemessungsansätze für das wirtschaftliche Verstärken von Betonbauteilen mit geklebter Bewehrung – Verbundtragfähigkeit unter statischer Belastung
Von *Konrad Zilch, Roland Niedermeier, Wolfgang Finckh.* 71,20 EUR

593: Praxisgerechte Bemessungsansätze für das wirtschaftliche Verstärken von Betonbauteilen mit geklebter Bewehrung – Verbundtragfähigkeit unter nicht ruhender Belastung (2013)
Von *Harald Budelmann, Thorsten Leusmann.* 44,50 EUR

594: Praxisgerechte Bemessungsansätze für das wirtschaftliche Verstärken von Betonbauteilen mit geklebter Bewehrung – Querkrafttragfähigkeit
Von *Konrad Zilch, Roland Niedermeier, Wolfgang Finckh.* 50,40 EUR

595: Erläuterungen und Beispiele zur DAfStb-Richtlinie „Verstärken von Betonbauteilen mit geklebter Bewehrung" (2013)
Von *Konrad Zilch.* 50,80 EUR

595 (en): Commentary on the DAfStb Guideline "Strengthening of concrete members with adhesively bonded reinforcement" with Examples (2014) 63,50 EUR

596: Vereinfachtes Rechenverfahren zum Nachweis des konstruktiven Brandschutzes bei Stahlbeton-Kragstützen (2013).
Von *Dietmar Hosser, Ekkehard Richter.* 25,60 EUR

597: Erweiterte Datenbanken zur Überprüfung der Querkraftbemessung für Konstruktionsbetonbauteile mit und ohne Bügel (2012).
Von *Karl-Heinz Reineck, Daniel A. Kuchma, Birol Fitik.* 192,40 EUR

598: Mischungsentwurf und Fließeigenschaften von Selbstverdichtendem Beton (SVB) vom Mehlkorntyp unter Berücksichtigung der granulometrischen Eigenschaften der Gesteinskörnung (2012).
Von *Andreas Huß.* 57,30 EUR

599: Bewehren nach Eurocode 2 (2013).
Von *Josef Hegger, Martin Empelmann, Jürgen Schnell, Jörg Moersch, Christian Albrecht, Guido Bertram, Norbert Brauer, Thomas Sippel, Marco Wichers.* 98,80 EUR

600: Teil 1: Erläuterungen zu DIN EN 1992-1-1 und DIN EN 1992-1-1/NA 2. überarbeitete Auflage (2020). 98,80 EUR

601: Dauerhaftigkeitsbemessung von Stahlbetonbauteilen auf Bewehrungskorrosion – Teil 1: Systemparameter der Bewehrungskorrosion (2012).
Von *Peter Schießl, Kai Osterminski, Bernd Isecke, Matthias Beck, Andreas Burkert, Jens Lehmann, Armin Faulhaber, Michael Raupach, Jörg Harnisch, Jürgen Warkus, Wei Tian, Christoph Gehlen.* 50,80 EUR

602: Dauerhaftigkeitsbemessung von Stahlbetonbauteilen auf Bewehrungskorrosion – Teil 2: Dauerhaftigkeitsbemessung (2012).
Von *Harald S. Müller, Edgar Bohner, Christian Fischer, Joško Ožbolt, Christoph Gehlen, Kai Osterminski, Peter Schießl, Stefanie von Greve-Dierfeld.* 68,60 EUR

603: Gütebewertung qualitativer Prüfaufgaben in der zerstörungsfreien Prüfung im Bauwesen am Beispiel des Impulsradarverfahrens (2012).
Von *Sascha Feistkorn.* 70,80 EUR

604: Frostbeanspruchung und Feuchtehaushalt in Betonbauwerken (2013).
Von *Frank Spörel.* 158,40 EUR

605: Zur Rheologie und den physikalischen Wechselwirkungen bei Zementsuspensionen (2012).
Von *Michael Haist.* 78,50 EUR

606: Unbewehrte Betonfahrbahnplatten unter witterungsbedingten Beanspruchungen (2014).
Von *Sam Foos.* 111,40 EUR

607: Modell zur Beschreibung des Eindringens von Chlorid in Beton von Verkehrsbauten (2013).
Von *Gesa Kapteina.* 63,90 EUR

608: Auswirkungen der Bewehrungskorrosion auf den Verbund zwischen Stahl und Beton (2013).
Von *Christian Fischer.* 58,20 EUR

609: Untersuchungen zum Verbundverhalten von Bewehrungsstäben mittels vereinfachter Versuchskörper (2013).
Von *Anke Wildermuth.* 132,60 EUR

610: Einfluss der Bauteilgeometrie auf die Korrosionsgeschwindigkeit von Stahl in Beton bei Makroelementbildung (2014).
Von *Jürgen Warkus.* 113,60 EUR

611: Sedimentationsverhalten und Robustheit Selbstverdichtender Betone (2014).
Von *Dirk Lowke.* 93,60 EUR

612: Bestimmung und Bewertung des elektrischen Widerstands von Beton mit geophysikalischen Verfahren (2014).
Von *Kenji Reichling* 94,00 EUR

613: Untersuchungen zur Leistungsfähigkeit von nationalen und europäischen Instandsetzungsmörteln (2015).
Von *Wolfgang Breit, Joachim Schulze* und *Delphine Schwab* 49,30 EUR

614: Erläuterungen zur DAfStb-Richtlinie Stahlfaserbeton (2015). 38,30 EUR

614: (en): Commentary on the DAfStb Guideline „Steel Fibre Reinforced Concrete"(2015) 47,90 EUR

615: Erläuterungen zu DIN EN 1992-4 – Bemessung der Verankerung von Befestigungen in Beton. 149,80 EUR

615 (en): Commentary to EN 1992-4 – Design of Fastenings for use in Concrete 149,80 EUR

616: Sachstandbericht Bauen im Bestand – Teil I: Mechanische Kennwerte historischer Betone, Betonstähle und Spannstähle für die Nachrechnung von bestehenden Bauwerken.
Von *Jürgen Schnell, Konrad Zilch, Daniel Dunkelberg und Michael Weber.* 98,80 EUR

617 (en): ACI-DAfStb databases 2015 with shear tests for evaluating relationships for the shear design of structural concrete members without and with stirrups.
Von *Karl-Heinz Reineck, Daniel Dunkelberg.* 292,90 EUR

618: Sachstandbericht - Grenzzustände der Ermüdung von dynamisch hoch beanspruchten Tragwerken aus Beton.
Von *Jürgen Grünberg, Michael Hansen, Steffen Marx, Sebastian Schneider.* 72,40 EUR

619: Sachstandsbericht Bauen im Bestand – Teil II: Bestimmung charakteristischer Betondruckfestigkeiten und abgeleiteter Kenngrößen im Bestand
Von *Jürgen Schnell, Konrad Zilch, Daniel Dunkelberg und Michael Weber.* 61,65 EUR

620: Sachstandbericht
Verfahren zur Prüfung des Säurewiderstands von Beton.
Von *Jesko Gerlach und Ludger Lohaus.* 51,60 EUR

621: Zur Verwertbarkeit von Potentialfeldmessungen für die Zustandserfassung und -prognose von Stahlbetonbauteilen – Validierung und Einsatz im Lebensdauermanagement.
Von *Sylvia Keßler.* 82,40 EUR

622: Bemessungsregeln zur Sicherstellung der Dauerhaftigkeit XC-exponierter Stahlbetonbauteile.
Von *Stefanie Marilies von Greve-Dierfeld.* 113,40 EUR

623: Untersuchungen an 43 Jahre im Nordseeklima ausgelagerten Betonbalken.
Von *Kai Osterminski und Christoph Gehlen.*
Bemessung auf Dauerhaftigkeit mit Teilsicherheitsbeiwerten und mit qualifiziert abgesicherten deskriptiven Regeln.
Von *Stefanie Marilies von Greve-Dierfeld und Christoph Gehlen.* 74,85 EUR

624: Stoffgesetz zur Beschreibung des Kriech- und Relaxationsverhaltens junger normal- und hochfester Betone.
Von *Isabel Anders.* 81,70 EUR

625: Zum Querkrafttragverhalten von einachsig gespannten Stahlbetonplatten ohne Querkraftbewehrung unter Einzellasten.
Von *Karin Reißen.* 112,90 EUR

626: Semiprobabilistisches Nachweiskonzept zur Dauerhaftigkeitsbemessung und -bewertung von Stahlbetonbauteilen unter Chlorideinwirkung.
Von *Amir Rahimi.* 116,20 EUR

627 (en): Shear Strength Models for Reinforced and Prestressed Concrete Members.
Von *Martin Herbrand.* 72,30 EUR

Heft

628: Zur Eignung und Wirkungsweise calcinierter Tone als reaktive Bindemittelkomponente im Zement.
Von *Nancy Beuntner.* 61,60 EUR

629: Zur einheitlichen Bemessung gegen Durchstanzen in Flachdecken und Fundamenten.
Von *Carsten Siburg.* 132,20 EUR

630: Bemessung nach DIN EN 1992 in den Grenzzuständen der Tragfähigkeit und der Gebrauchstauglichkeit.
98,80 EUR

631: Hilfsmittel zur Schnittgrößenermittlung und zu besonderen Detailnachweisen bei Stahlbetontragwerken. 89,90 EUR

632: Experimentelle Ermittlung der Korrelation der Druckfestigkeiten von Bohrkernen aus Bauwerksbeton und genormten Probekörpern.
Von *Jürgen Schnell und Michael Weber.* 62,37 EUR

633: (en): Steel Fiber Reinforced Concrete Under Concentrated Load.
Von *Fanbing Song.* 120,15 EUR

634: Vergleichende Untersuchungen zur Rückprallhammerprüfung bezogen auf R- und Q-Werte.
Von *Wolfgang Breit und Melanie Merkel.* 27,50 EUR

635-1 (en): ACI-DAfStb databases 2020 with shear tests on structural concrete members without stirrups Volume 1: Part 1 to Part 2.5.
Von *Karl-Heinz Reineck und Birol Fitik.*
190,40 EUR

635-2 (en): ACI-DAfStb databases 2020 with shear tests on structural concrete members without stirrups Volume 2: Part 2.6 to Part 7.
Von *Karl-Heinz Reineck und Birol Fitik.*
237,30 EUR

636: Untersuchungen zu Anforderungen an die Dauerhaftigkeit von R4-Produkten nach EN 1504-3 – Ringversuch zur Ermittlung des Chlorideindringwiderstands von Mörtel und Beton.
Von *Wolfgang Breit, Joachim Schulze*
84,81 EUR

637: Sachstandbericht Frischbeton – Eigenschaften, Einflüsse und Prüfungen.
Von *Christoph Alfes, Olaf Aßbrock, Christoph Begemann, Rolf Breitenbücher, Amela Cokovik, Dario Cotardo, Eberhard Eickschen, Petra Fischer, Eugen Kleen, Hannes Krüger, Ludger Lohaus, Viktor Mechtcherine, Lars Meyer, Matthias Middel, Christoph Müller, Harald S. Müller, Tobias Schack, Patrick Schäffel, Egor Secrieru, Frank Spörel, Katja Voland, Andreas Westendarp und Bou-Young Youn-Ĉale* 86,40 EUR

Heft

638: Anwendungshilfe zur Technischen Regel Instandhaltung von Betonbauwerken des DIBt (TR IH) in Verbindung mit der DAfStb Richtlinie Schutz und Instandsetzung von Betonbauteilen (RL SIB). 115,00 EUR

639: Schlussberichte zum BMBF-Verbundforschungsvorhaben „R-Beton – Ressourcenschonender Beton – Werkstoff der nächsten Generation“ Schwerpunkt 1: Konzeptionierung der neuen Werkstoffe.
Von *Florian Knappe, Joachim Reinhardt, Stefanie Theis, Stephan Kresser, Julia Scheidt, Wolfgang Breit, Bernard Sachsenhauser, Klaus Lorenz, Christoph Müller, Katrin Severins, Johannes Haufe und Anya Vollpracht* 103,30 EUR

640: Schlussberichte zum BMBF-Verbundforschungsvorhaben „R-Beton – Ressourcenschonender Beton – Werkstoff der nächsten Generation“ Schwerpunkt 2: Praxisanforderungen an die neuen Werkstoffe.
Von *Aleksandar Pančić, Jürgen Schnell, Christoph Müller, Ingmar Borchers, Maik Seidel und Anya Vollpracht, Lia Weiler* 95,00 EUR

641: Schlussberichte zum BMBF-Verbundforschungs vorhaben „R-Beton – Ressourcenschonender Beton – Werkstoff der nächsten Generation“ Schwerpunkt 3: Ökobilanz, Praxistest und Transfer.
Von *Florian Knappe, Joachim Reinhardt, Stefanie Theis, Christoph Müller, Jochen Reiners und Raymund Böing*
75,80 EUR

642 (en): Influence of elevated temperatures up to 100 °C on the mechanical properties of concrete.
Von *Fernando Acosta Urrea*
88,00 EUR

643 Realkalisierungspotenzial von zementgebundenen Werkstoffen im Trinkwasserbereich
Von *Melanie Merkel* 76,40 EUR

644: Ermüdungsverhalten von Beton für unterschiedliche Probekörpergeometrien.
Von *Vivian Frei, Stephan Pirskawetz, Marc Thiele, Andreas Rogge*
68,80 EUR

645: Modellhafte Beschreibung des Ermüdungswiderstands von druckschwellbeanspruchtem Beton unter Berücksichtigung von energetischen und frequenzbedingten Materialeffekten.
Von *Sebastian Schneider, Matthias Bode, Steffen Marx* 63,20 EUR

646: Wasserinduzierte Ermüdungsschädigung von Beton
Von *Christoph Tomann, Ludger Lohaus*
84,40 EUR

647: Untersuchungen zum Ermüdungswiderstand von Beton im Bereich sehr hoher Lastwechselzahlen.
Von *Kerstin Willers, Lutz Gerlach, Nico Herrmann* 53,30 EUR

Heft

648: Zum festigkeitsabhängigen Ermüdungswiderstand von Beton. Teil 1: Koordinierte experimentelle Untersuchungen und Bewertung des Ermüdungswiderstands von normal- und hochfesten Betonen.
Von *Sebastian Schneider, Boso Schmidt, Steffen Marx*
Teil 2: Statistische Auswertungen zum Druckfestigkeitseinfluss auf den Ermüdungswiderstand von Beton.
Von *Marco Basaldella, Bianca Kern, Nadja Oneschkow, Ludger Lohaus*
52,80 EUR

649: Numerische Modellierung des Ermüdungsverhaltens von normal- und hochfestem Beton.
Von *Dennis Birkner, Steffen Marx, Abedulgader Baktheer, Josef Hegger, Rostislav Chudoba* 58,40 EUR

650: Ermüdung von biegebeanspruchten Betonbauteilen aus normal- und hochfesten Betonen.
Von *Dennis Birkner, Steffen Marx, David Ov, Rolf Breitenbücher*
40,50 EUR

651: Ultraschallprüfungen zur Erfassung der Schädigungsentwicklung unter verschiedenen Umweltbedingungen und unter zyklischer Druckschwellbelastung.
Von *Raúl Beltrán, Vivian Frei, Steffen Marx* 56,60 EUR

652: Ermüdungsverhalten von Betonstahl im Langzeitfestigkeitsbereich, bei Variation der Prüfmethodik sowie unter kombinierter Einwirkung von Korrosion.
Von *Stefan Rappl, Kai Osterminski, Christoph Gehlen* 37,70 EUR

653: Verbundverhalten unter Druck- und Zugschwellbeanspruchung von normal- und hochfesten Betonen
Von *Marc Koschemann, Manfred Curbach, Homam Spartali, Abedulgader Baktheer, Rostislav Chudoba, Josef Hegger* 54,20 EUR

654: Zeitlich und räumlich aufgelöste Zustandsbewertungen und -prognosen über mit Diagnosedaten angereicherte BIM-Modelle als Entscheidungshilfe bei der Instandsetzungsplanung
Von *Hendrik W. L. Morgenstern*
126,80 EUR

656: Ermittlung von Zwangschnittgrößen in Hochbaudecken aus Stahlbeton
Von *Andreas Dridiger, Andrej Albert, Jürgen Schnell* 128,70 EUR

657 (en): Impact of Microstructural Changes in Concrete on Coda Waves
Von *Fabian Diewald* ca. 88,00 EUR

658: Zur Deckschichtbildung an feuerverzinkten Betonstählen in chloridhaltigen Mörteln
Von *Gino Ebell* 87,20 EUR

660: Erläuterungen und Beispiele zur DAfStb-Richtlinie „Betonbauteile mit nichtmetallischer Bewehrung“
439,00 EUR

Hinweis auf überarbeitete und ergänzte Hefte der Schriftenreihe des DAfStb:

Heft 220: 2. überarbeitete Auflage 1991
Heft 240: 3. überarbeitete Auflage 1991 (vergriffen)
Heft 400: 4. Auflage 1994 (3. berichtigter Nachdruck) vergriffen
Heft 425: 3. ergänzte Auflage 1997
Heft 525: 2. überarbeitete Auflage 2010
Heft 600: 2. überarbeitete Auflage 2020

DAfStb-Heft 657 (en)